ALGEBRA

A MODERN APPROACH

ABOUT THE AUTHOR

Kaj L. Nielsen received the degrees of B.S. from the University of Michigan, M.A. from Syracuse University, and Ph.D. from the University of Illinois. He has held teaching positions at Syracuse, Illinois, Brown, Butler, Purdue, and Louisiana State University. He has been associated with a number of industries and the U. S. Navy Department as a research engineer and scientist. He is presently Director of the Systems Analysis Division at Battelle Memorial Institute in Columbus, Ohio. Dr. Nielsen has written numerous articles based upon original research in mathematics and published in leading mathematical and engineering journals. He is the author or coauthor of *College Mathematics, Modern Trigonometry, Differential Equations, Logarithmic and Trigonometric Tables, Plane and Spherical Trigonometry,* and *Problems in Plane Geometry,* in the College Outline Series, and of *Mathematics for Practical Use,* in the Everyday Handbooks Series. Dr. Nielsen is listed in *American Men of Science, Who is Who in the Midwest, Who is Who in the Computer Field,* and *The World Directory of Mathematicians.*

COLLEGE OUTLINE SERIES

ALGEBRA

A MODERN APPROACH

KAJ L. NIELSEN

BARNES & NOBLE BOOKS

A DIVISION OF HARPER & ROW, PUBLISHERS

New York, Hagerstown, San Francisco, London

To
Cheryl

GREEK ALPHABET

Letters		Names	Letters		Names	Letters		Names
A	α	Alpha	I	ι	Iota	P	ρ	Rho
B	β	Beta	K	κ	Kappa	Σ	σ	Sigma
Γ	γ	Gamma	Λ	λ	Lambda	T	τ	Tau
Δ	δ	Delta	M	μ	Mu	Υ	υ	Upsilon
E	ϵ	Epsilon	N	ν	Nu	Φ	ϕ	Phi
Z	ζ	Zeta	Ξ	ξ	Xi	X	χ	Chi
H	η	Eta	O	o	Omicron	Ψ	ψ	Psi
Θ	θ	Theta	Π	π	Pi	Ω	ω	Omega

SYMBOLS

The language of mathematics makes extensive use of symbolic notation as a means of communication. The modern trend is to expand this symbolic language, and the glossary of nomenclature has grown quite large. Fortunately, once a symbol has been accepted, it becomes a part of a universal language. In order to focus attention upon the more important symbols of algebra, we present a summary of symbols at the end of each chapter.

PREFACE

This book presents the fundamentals of a first course in college algebra. In its development the text follows what has been called the modern approach: it bases the logical structure of algebra on the set concepts. Thus it will furnish both a rapid review for the reader who has completed this course of study and supplementary reading for one who is enrolled in the course. Designed to be used as a self-tutor, it is written so that a conscientious reader can become proficient in this subject by studying thoughtfully, practicing the appropriate techniques, and working the exercises.

A course in college algebra may include many things. Every instructor likes to emphasize particular points. This book concentrates on fundamental concepts in order to satisfy the needs of the majority of students. The book is complete in itself, and although it is not intended to be used as a textbook, it may be read with profit by those who approach modern algebra for the first time. It should provide any reader with a thorough knowledge of the usual topics studied in this very interesting field of mathematics.

Considerable space is devoted to the development of the number system from the set-theoretic viewpoint. Algebra is based on the number system, and set theory provides powerful and rigorous means for developing that system. The more common algebraic tools are illustrated throughout the book so that the reader can acquire the skills of computational algebra as well as an understanding of the basic logic behind mathematical structures. Although the subject matter is developed in the modern style, the classical aspects of algebra are also covered and the thorough reader should not fear that necessary algebraic techniques have been omitted.

Each concept is supported by figures, solved problems, and

examples. For purposes of clarity, we make a distinction between examples and illustrations; an example is used to exhibit typical elements of an idea, whereas an illustration states a problem and displays the solution to that problem. Neatness and the employment of a systematic manner of solving problems, both of which are valuable in the study of any branch of mathematics, have been stressed. The use and adaptation of systematic procedures contribute significantly to the development of logical thinking and the elimination of errors.

Abbreviated tables have been placed in the back of the book for computational purposes. If the reader finds a need for greater accuracy, he can obtain more extensive tables in the College Outline Series.

Thorough knowledge of a branch of mathematics cannot be obtained without solving problems. Consequently, some typical exercises have been placed at judicious intervals throughout the book. The student is urged to work all the exercises. Answers are furnished for those that do not involve lengthy proofs or diagrams, so that the reader may check his work. A set of examinations, with answers, is also provided. These sample tests should aid the student in his preparation for any examination.

Since the modern revolution in mathematics has caused many topics to be moved from college to advanced high schools, many honors students will have covered some (perhaps all) of the topics in this book. It has been the experience of the author that such students frequently need a quick reference to supplement their subsequent studies. This book is intended also to satisfy that need.

No single book in mathematics can cover all topics in any given field, and this book is no exception. It is hoped that the reader will not consider this book as a summary of or an end to his study of mathematics, but rather as a sourcebook and reference which will lead into other mathematical pursuits.

The author gratefully acknowledges his indebtedness to his many students, who, during his teaching career, contributed so much to the manner of his presentation of college mathematics; to the scholars who developed the subject matter; to his daughter Cheryl, who asked many pertinent questions while she was studying the same material; to his friends who aided in the preparation of the manuscript; and to the staff of Barnes & Noble, Inc., for their usual pleasant cooperation and their excellent contributions to the finished product.

Kaj L. Nielsen

TABLE OF CONTENTS

I

REVIEW

1. Introduction. Mathematics is a branch of knowledge which involves a great many topics. In the elementary stages of the subject, the topics fall into a logical sequence of development. We are about to study the topics of introductory modern algebra and we must presume a certain amount of mathematical background. In this chapter we shall consider the material with which the reader should already be familiar. Consequently, this first chapter is in the nature of a review and the material is presented only for easy reference. Should the reader desire more detailed explanations of these concepts, he might consult the author's book *Mathematics For Practical Use*, New York: Barnes and Noble, Inc., 1962.

2. The Numbers. One of the first objects of study in mathematics is the numbers. In our study of arithmetic we were introduced to the integers, fractions, and decimals. We should consequently be familiar with

    positive integers: 1, 2, 3, 4, 5, . . .
    zero integer: 0
    negative integers: $-1, -2, -3, -4, \ldots$
    fractions: $1/2, 2/3, 7/8, 15/4, -3/4, \ldots$
    decimal fractions: $.25, .01, 1.0024, -2.31, \ldots$

We should also be familiar with some of the names and concepts concerning numbers.

An **even number** is an integer that is exactly divisible by 2.
Examples: 2, 4, 6, 8, 24, 256
An **odd number** is an integer that is not exactly divisible by 2.
Examples: 1, 3, 5, 7, 33, 51, 1143
A **factor** of a number is an integer that divides it exactly.
Examples: 3 and 5 are factors of 15; 2, 4, 5, and 10 are factors of 20.
A **prime number** is an integer which has no factors except 1 and itself.
Examples: 2, 7, 11, 13

A **composite number** is a number that is not prime.

Examples: 4, 22, 35, 1028

A **common divisor** of two or more numbers is a factor that will exactly divide each of them.

Example: 3 is a common divisor of 9 and 27.

The **greatest common divisor** (**G.C.D.**) is the largest factor of two or more numbers.

Example: 9 is the G.C.D. of 27 and 45.

The **absolute** or **numerical value** of a number is the magnitude of the number and is denoted by two vertical bars, $|n|$.

Examples: $|5| = 5$ and $|-5| = 5$.

The arithmetic of numbers should be familiar to the student. We shall briefly summarize a few operations which seem to be troublesome.

1. To add or subtract two or more fractions, find the **lowest common denominator** (L.C.D.), change each fraction to its equivalent over the L.C.D., and add the numerators.

Example:

$$\frac{2}{3} + \frac{3}{4} - \frac{5}{6} = \frac{2(4) + 3(3) - 5(2)}{12} = \frac{8 + 9 - 10}{12} = \frac{7}{12}$$

2. To multiply two or more fractions, multiply numerator by numerator and denominator by denominator; reduce the result to its lowest form.

Example:

$$\frac{2}{3} \times \frac{3}{4} \times \frac{5}{6} = \frac{2 \times 3 \times 5}{3 \times 4 \times 6} = \frac{30}{72} = \frac{5}{12}$$

or

$$\frac{2}{3} \times \frac{3}{4} \times \frac{5}{6} = \frac{\overset{1}{\cancel{2}} \times \overset{1}{\cancel{3}} \times 5}{\underset{1}{\cancel{3}} \times \underset{2}{\cancel{4}} \times 6} = \frac{1 \times 1 \times 5}{1 \times 2 \times 6} = \frac{5}{12}$$

3. To divide two fractions, multiply the numerator by the reciprocal of the denominator.

Example:

$$\frac{2}{3} \div \frac{5}{7} = \frac{2}{3} \times \frac{7}{5} = \frac{14}{15}$$

4. To add numbers containing decimals, add the corresponding digits, retaining the position of the decimal point.

Illustration. Find the sum of $1.23 + 6.5 + 0.149$.

Solution.

$$\begin{array}{r} 1.230 \\ 6.500 \\ 0.149 \\ \hline 7.879 \end{array}$$

5. To multiply numbers containing decimals, multiply each digit of the multiplicand by each digit of the multiplier and add the number of decimal places to obtain the position of the decimal point in the product.

Illustration. Multiply 3.14 by 2.16.
Solution.

$$\begin{array}{r} 3.14 \quad \textbf{(multiplicand)} \\ 2.16 \quad \textbf{(multiplier)} \\ \hline 1\,8\,8\,4 \\ 3\,1\,4 \\ 6\,2\,8 \\ \hline 6.7\,8\,2\,4 = 6.78 \quad \textbf{(product)} \end{array}$$

6. To obtain a given per cent of a number, change the per cent to its decimal equivalent and multiply.

Illustration. What is $2\frac{1}{2}\%$ of 325.00?
Solution.

$$2\frac{1}{2}\% = .025$$
$$N = (.025)(325.00) = 8.125$$

The student should review his multiplication tables, the powers of the integers, and the roots of the integers. Table I in the back of this book will be useful. The advantage of memorizing the following table will become apparent throughout the study of the book.

POWER N	2	3	4	5	6	7	8	9
2	4	8	16	32	64	128	256	512
3	9	27	81	243	729			
4	16	64	256	1024				
5	25	125	625					
6	36	216						

3. Symbols and Signs. A part of mathematics is a shorthand of symbols and signs which we must learn as a part of the language.

The basic ones are

I. *The signs of operations*
   a) Addition: + ; e.g., 3 + 2
   b) Subtraction: − ; e.g., 5 − 3
   c) Multiplication: × or · ; e.g., 7 × 3 or 7 · 3
   d) Division: ÷ ; e.g., 4 ÷ 3

II. *The signs of order*
   a) Is less than: < ; e.g., 4 < 9
   b) Is greater than: > ; e.g., 7 > 5
   c) Is equal to: = ; e.g., 7 = (14 ÷ 2)

III. *The signs of grouping*
   a) The parentheses: ( ) ; e.g., (3 + 2) + (4 + 7)
   b) The brackets: [ ] ; e.g., [(3 + 2) + 5] − 6
   c) The braces: { } ; e.g., {[8 + (3 − 1)] − (6 + 2)}

In the common fraction, the line denotes a division; for example, $\frac{2}{3}$ or 2/3 means 2 divided by 3. The parenthesis is also used to denote a multiplication; for example, $(2 + 3)(7 − 4)$ means $(2 + 3) \times (7 − 4)$.

We use a minus sign, "−," to indicate that a number is negative and sometimes a plus sign, "+," to indicate that a number is positive. Thus both signs have a double meaning; i.e., a plus sign may
   a) indicate an operation (to add),
   b) indicate a positive number;
a minus sign may
   a) indicate an operation (to subtract),
   b) indicate a negative number.

The discussion in which these symbols are used invariably makes it clear which of the two meanings is intended. We shall encounter many more symbols in the course of reading this book, for mathematics makes extensive use of a universally accepted shorthand notation.

4. The Formula.   The area of a rectangle is obtained by multiplying the length by the width. In the notation of mathematics, this fact is written $A = ab$, where A represents the area, a represents the length, b represents the width, and ab means a times b. This is a simple introduction to the algebraic formula, which expresses a fact. Letters are used to represent numbers. The letters have different values depending upon the problem under discussion and their meanings must be clearly defined. They are frequently referred to as **general numbers.** The study of algebra is essentially concerned with the operations of, establishing laws for, and understanding the meaning of such general numbers.

Examples:

$$\text{Area of a circle:} \quad A = \pi r^2$$
$$\text{Area of a triangle:} \quad A = \tfrac{1}{2}bh$$

Furthermore, we shall be concerned with combinations of symbols. If each symbol in a combination represents a number, we shall call the combination an **expression.**
Examples:

$$2ax, \quad 2x - 3a, \quad 4y^2 + 3ax - 12$$

If there are several parts connected by plus and minus signs, each part is called a **term.** Thus, in the above examples, $2x$ is a term and $-3a$ is a term. Note that the sign is associated with the term, but if the sign is $+$ it is usually not expressed except for emphasis.

In discussing a term of an expression we have additional terminology. Thus we say that each of two or more quantities that are multiplied together is a **factor** of the product. Furthermore, any factor of an expression is called the **coefficient** of the remaining part.
Example:

$$32bx^2$$

32, b, and x^2 are factors.
32 is the numerical coefficient
$32b$ is the coefficient of x^2.
$32x^2$ is the coefficient of b.

Names are also given to various expressions. Thus we have

monomial, an expression consisting of one term,
binomial, an expression consisting of two terms,
trinomial, an expression consisting of three terms,
polynomial, an expression consisting of more than one term.

5. Fundamental Operations. It is necessary that we establish some ground rules in order that we may proceed in an orderly manner. We start with the four fundamental operations of addition, subtraction, multiplication, and division. First we have three important laws:
I. **The Commutative Law.** *The result of addition or multiplication is the same in whatever order the terms are added or multiplied.*

$$a + b = b + a \quad \text{and} \quad ab = ba$$

II. **The Associative Law.** *The sum of three or more terms, or the product of three or more factors, is the same in whatever manner they are grouped.*

$$a + (b + c) = (a + b) + c = a + b + c$$
$$a(bc) = (ab)c = abc$$

III. The Distributive Law. *The product of an expression of two or more terms by a single factor is equal to the sum of the products of each term of the expression by the single factor.*

$$(a + b - c)d = ad + bd - cd$$

Next, we have some **laws of signs:**

I. To add two numbers having like signs, *add their absolute values and prefix the common sign.*
Examples:

$$3 + 5 = 8 \qquad -3 + (-5) = -8$$

II. To add two numbers having unlike signs, *take the difference of their absolute values and prefix to it the sign of the number having the larger absolute value.*
Examples:

$$5 + (-2) = 3 \qquad 2 + (-5) = -3$$

III. To subtract one number from another, *change the sign of the number to be subtracted and proceed as in addition.*
Examples:

$$7 - (+3) = 7 + (-3) = 4$$
$$3 - (+7) = 3 + (-7) = -4$$
$$3 - (-7) = 3 + (+7) = 10$$

IV. The product, or quotient, of two numbers *has the following rule of signs:*

> *a* **positive** *by a* **positive** *yields a* **positive;**
> *a* **positive** *by a* **negative** *yields a* **negative;**
> *a* **negative** *by a* **positive** *yields a* **negative;**
> *a* **negative** *by a* **negative** *yields a* **positive.**

Examples:

$$3 \times 2 = 6 \qquad 6 \div (-2) = -3$$
$$-4 \times 2 = -8 \qquad (-8) \div (-4) = 2$$

Symbols of grouping preceded by a plus sign *may be removed by rewriting each of the inclosed terms with its original sign.*
Example:

$$3x + (2y - 4z) = 3x + 2y - 4z$$

Symbols of grouping preceded by a minus sign *may be removed by rewriting and changing the sign of each of the inclosed terms.*
Example:

$$3x - (2y - 4z) = 3x - 2y + 4z$$

If a **coefficient** *precedes a symbol of grouping, it is to be multiplied into each included term when the symbol is removed.*
Example:

$$x - 2(3y - 2a) = x - 6y + 4a$$

To simplify expressions involving several symbols of grouping we work from the inside out by *first removing the innermost pair* of symbols, next the innermost pair of the remaining ones, and so on. If like terms appear they are combined at each step as in the following example.
Example:

$$
\begin{aligned}
4a - b - &\{3a - [2a(4 - b) - (a - b)]\} \\
&= 4a - b - \{3a - [8a - 2ab - a + b]\} \\
&= 4a - b - \{3a - 7a + 2ab - b\} \\
&= 4a - b + 4a - 2ab + b \\
&= 8a - 2ab
\end{aligned}
$$

6. Powers, Exponents, and Radicals. In the case of multiplying together factors that are all alike, a shorthand method of writing the product has been devised, and the product is called a **power** of the factor. Thus, $x \cdot x \cdot x$ is the *third* power of x and we write it as x^3. The number x is called the **base** and 3 the **exponent** of the power. Thus the exponent is the number of times the like factor appears in the multiplication. We speak of

$$
\begin{array}{lll}
x^2 & \text{as} & \text{``}x \text{ square,''} \\
x^3 & \text{as} & \text{``}x \text{ cube,''} \\
x^n & \text{as} & \text{``the } n^{th} \text{ power of } x.\text{''}
\end{array}
$$

Examples:

$$(-2)(-2)(-2)(-2) = (-2)^4 = 16$$
$$x^5 = x \cdot x \cdot x \cdot x \cdot x = \text{five factors of } x$$

In the above discussion we have limited ourselves to the case where the exponent is a positive integer. We shall generalize this, but before we do so, let us consider the **basic laws of exponents.** In the following, let m and n be any two positive integers.

$$\text{I.} \quad a^m \cdot a^n = a^{m+n}$$

$$\text{II.} \quad (a^m)^n = a^{mn}$$

$$\text{III.} \quad (ab)^n = a^n b^n$$

$$\text{IV.} \quad \left(\frac{a}{b}\right)^n = \frac{a^n}{b^n}, \quad b \neq 0$$

$$\text{V.} \quad \frac{a^m}{a^n} = a^{m-n}, \quad m > n, \quad a \neq 0$$

$$\text{VI.} \quad \frac{a^m}{a^n} = \frac{1}{a^{n-m}}, \quad m < n, \quad a \neq 0$$

These laws may be proved by considering the definitions, e.g.:

$$a^m \cdot a^n = \underbrace{(a \cdot a \cdot a \ldots a)}_{m \text{ factors}} \underbrace{(a \cdot a \cdot a \ldots a)}_{n \text{ factors}}$$

$$= (a \cdot a \cdot a \ldots a), \quad m + n \text{ factors}$$

$$= a^{m+n}$$

The student may like to prove the others by writing out the definitions and collecting terms.

Examples:

$$2^3 \cdot 2^5 = 2^8 = 256$$

$$(3^2)^3 = 3^6 = 729$$

$$(2 \cdot 3)^2 = 2^2 \cdot 3^2 = 4 \cdot 9 = 36$$

$$\left(\frac{2}{3}\right)^2 = \frac{2^2}{3^2} = \frac{4}{9}$$

$$\frac{2^5}{2^2} = 2^3 = 8$$

$$\frac{3^2}{3^3} = \frac{1}{3}$$

The next logical step is to define a root, which we do by the statement **if $a^m = N$, then a is the mth root of N.** That is, we are looking for a number a such that when multiplied by itself to m factors it yields N. We write this statement using a **radical** sign, $\sqrt{}$, thus:

$$\sqrt[m]{N} = a$$

The process of taking a root does not necessarily lead to a single answer. For example, the square root of 4 may be either $+2$ or -2. This condition occurs for all even roots. However, the sign in front of the radical indicates which root is to be considered. If both roots are desired the sign $\pm$ (read "plus or minus") is used.

Examples:

$$\pm \sqrt{16} = \pm 4 \qquad \sqrt[4]{81} = 3 \qquad -\sqrt[4]{81} = -3$$

$$\sqrt[3]{-8} = -2 \qquad \sqrt[5]{32} = 2 \qquad \sqrt{a^2} = |a|$$

In the symbol $\sqrt[q]{N}$, q is called the **index of the root** and N, the **radicand.** The radicals obey the same laws as exponents, but before we turn to the proof we must first consider exponents which are not integers. We begin with the *fractional exponents.* Let $m = 1/q$, then according to Law I we may write

$$\underbrace{a^{\frac{1}{q}} \cdot a^{\frac{1}{q}} \ldots a^{\frac{1}{q}}}_{q \text{ factors}} = a^{\frac{q}{q}} = a$$

which is our definition of the qth root of a, and we have

$$a^{\frac{1}{q}} = \sqrt[q]{a}$$

If we raise this expression to the pth power, we write

$$\left(a^{\frac{1}{q}}\right)^p = a^{\frac{p}{q}} = (\sqrt[q]{a})^p$$

and we have the complete definition of a fractional exponent, which conforms to the laws of exponents.

Examples:

$$4^{\frac{1}{2}} = \sqrt{4} = 2 \qquad 27^{\frac{1}{3}} = 3$$

If we let $m = 0$, we have, according to Law I

$$a^0 \cdot a^n = a^{0+n} = a^n$$

or $$a^0 = \frac{a^n}{a^n} = 1, \quad a \neq 0$$

so that the definition is

$$a^0 = 1, \quad a \neq 0$$

Examples:

$$(3x)^0 = 1, \quad x \neq 0 \qquad 5y^0 = 5, \quad y \neq 0$$

If we let $m = -n$, where n is positive, we may write, according to Law I

$$a^{-n} \cdot a^n = a^{-n+n} = a^0 = 1, \quad a \neq 0$$

so that the definition reads

$$a^{-n} = \frac{1}{a^n}, \quad a \neq 0$$

Examples:

$$2x^2 y^{-3} = \frac{2x^2}{y^3} \qquad 3(xy)^{-2} = \frac{3}{x^2 y^2}$$

We have now defined exponents and roots for all practical purposes. The student should write down examples and have a thorough knowledge of these definitions before reading further. It should also be noted that the laws of exponents apply to radicals since radicals are simply fractional exponents. In particular, the following four laws may be emphasized (*a* and *b* are positive):

$$\text{I.} \quad (\sqrt[n]{a})^n = a$$

$$\text{II.} \quad (\sqrt[n]{a})(\sqrt[n]{b}) = \sqrt[n]{ab}$$

$$\text{III.} \quad \sqrt[m]{\sqrt[n]{a}} = \sqrt[mn]{a}$$

$$\text{IV.} \quad \frac{\sqrt[n]{a}}{\sqrt[n]{b}} = \sqrt[n]{\frac{a}{b}}$$

The simplification of radicals usually reduces to one of the following operations:

1. **Removing factors from the radicand.**
Example:

$$\sqrt{50x^3} = \sqrt{(25x^2)2x} = 5x\sqrt{2x}$$

2. **Introducing coefficients under the radical sign.**
Example:

$$2a\sqrt[3]{3b} = \sqrt[3]{(2a)^3}\,\sqrt[3]{3b} = \sqrt[3]{(8a^3)(3b)} = \sqrt[3]{24a^3 b}$$

3. **Reducing to a radical of lower order.**
Example:

$$\sqrt[6]{81} = \sqrt[6]{3^4} = 3^{\frac{4}{6}} = 3^{\frac{2}{3}} = \sqrt[3]{3^2} = \sqrt[3]{9}$$

4. **Rationalizing the denominator.**
Examples:

$$\sqrt{\frac{3}{5}} = \sqrt{\frac{3}{5} \cdot \frac{5}{5}} = \frac{\sqrt{15}}{\sqrt{25}} = \frac{1}{5}\sqrt{15}$$

$$\frac{\sqrt{2}}{\sqrt{7}} = \frac{\sqrt{2}}{\sqrt{7}} \cdot \frac{\sqrt{7}}{\sqrt{7}} = \frac{1}{7}\sqrt{14}$$

The above four operations on radicals will be used again in Section 11.

7. Multiplication and Division.   The multiplication of algebraic expressions is easily accomplished by paying strict attention to the laws of operations discussed in Section 5 and the laws governing exponents discussed in Section 6. Let us begin by operating on polynomials, which are expressions consisting of more than one term.

To find the **product of two polynomials,** *multiply each term of the* **multiplicand** *by each term of the* **multiplier** *and add the results.*

It is important to arrange the work systematically, and it is suggested that the multiplicand and multiplier be arranged in descending powers of one letter.

Illustration.   Multiply $(3x^2 - y^2 + 2xy)$ by $(y^2 - x^2 - 3xy)$.
Solution.

$$
\begin{array}{l}
3x^2 + 2xy - y^2 \qquad \textbf{(multiplicand)} \\
-x^2 - 3xy + y^2 \qquad \textbf{(multiplier)} \\
\hline
-3x^4 - 2x^3y + x^2y^2 \\
 - 9x^3y - 6x^2y^2 + 3xy^3 \\
 3x^2y^2 + 2xy^3 - y^4 \\
\hline
-3x^4 - 11x^3y - 2x^2y^2 + 5xy^3 - y^4 \quad \textbf{(product)}
\end{array}
$$

To check a multiplication, replace the letters by numbers in the multiplicand, multiplier, and product.

Check.   Let $x = 2$ and $y = 3$ in the above illustration.
Multiplicand $= 3(4) + 2(2)(3) - 9 = 15$
Multiplier $= -4 - 3(2)(3) + 9 = -13$
Product $= -3(16) - 11(8)(3) - 2(4)(9) + 5(2)(27) - 81 = -195$
Multiplicand $\times$ multiplier $= (15)(-13) = -195$

The operation of **division** is the inverse of multiplication. It is defined by the relation:

$$\textbf{Dividend} = \textbf{divisor} \times \textbf{quotient} + \textbf{remainder}$$

If the remainder is zero, the division is **exact** and the divisor is said to be a **factor** of the dividend. *To find the quotient of one polynomial divided by another* carry out the following steps:

1. Arrange each polynomial in descending powers of a common letter.

2. Divide the first term of the *dividend* by the first term of the *divisor,* yielding the first term of the *quotient.*

3. Multiply the *whole* divisor by the first term of the quotient, and subtract the product from the dividend.

4. The remainder is a new dividend, and steps 2 and 3 are repeated on this dividend.

5. Continue the process until a remainder is obtained which is an expression whose first term does not contain the first term of the divisor as a factor.

6. Check the result by replacing the letters by numbers. Do not use numbers which make the divisor zero.

The work should be arranged schematically; a scheme is suggested in the following illustration.

Illustration. Divide $15x^3 + 23x - 26x^2 - 3$ by $3x^2 + 3 - 4x$.

Solution.

$$
\begin{array}{c|l}
\textbf{(dividend)} \quad 15x^3 - 26x^2 + 23x - 3 & 3x^2 - 4x + 3 \quad \textbf{(divisor)} \\
\phantom{\textbf{(dividend)} \quad} 15x^3 - 20x^2 + 15x & 5x \;\; - 2 \quad\quad \textbf{(quotient)} \\
\cline{1-1}
 - 6x^2 + 8x - 3 & \\
 - 6x^2 + 8x - 6 & \\
\cline{1-1}
 3 \quad \textbf{(remainder)} &
\end{array}
$$

Check. Let $x = 2$.

Dividend $= 15(8) - 26(4) + 23(2) - 3 = 59$

Divisor $= 3(4) - 4(2) + 3 = 7$

Quotient $= 5(2) - 2 = 8$

Remainder $= 3$

Dividend $=$ divisor $\times$ quotient $+$ remainder

$59 = (7)(8) + 3 = 56 + 3 = 59$

8. Factoring. We shall now consider the process of finding two or more expressions whose product is a given expression. This process is called **factoring** and is in reality exact division or the inverse of multiplication. We accomplish this by being able to recognize a group of typical products, and upon seeing them in an algebraic expression we can immediately write down the factors. The first step, therefore, is to memorize the following products (the student may verify them by multiplying the right-hand factors):

1. $ab + ac = a(b + c)$
2. $a^2 + 2ab + b^2 = (a + b)^2$
3. $a^2 - 2ab + b^2 = (a - b)^2$
4. $a^2 - b^2 = (a + b)(a - b)$
5. $x^2 + (a + b)x + ab = (x + a)(x + b)$
6. $a^3 + b^3 = (a + b)(a^2 - ab + b^2)$
7. $a^3 - b^3 = (a - b)(a^2 + ab + b^2)$
8. $acx^2 + (ad + bc)x + bd = (ax + b)(cx + d)$

9. $a^2 + b^2 + c^2 + 2ab + 2ac + 2bc = (a + b + c)^2$

10. $a^3 + 3a^2b + 3ab^2 + b^3 = (a + b)^3$

With these basic products firmly in mind, it is possible to factor algebraic expressions by inspection. To ease the process of factoring, first remove all common expressions and then make advantageous groupings.

Examples:

a) $\begin{aligned} 27x^2 + 36xy + 12y^2 &= 3(9x^2 + 12xy + 4y^2) \\ &= 3(3x + 2y)^2 \end{aligned}$

b) $\begin{aligned} ax^2 - ay^2 - 6ay - 9a &= a(x^2 - y^2 - 6y - 9) \\ &= a[x^2 - (y^2 + 6y + 9)] \\ &= a[x^2 - (y + 3)^2] \\ &= a[x - (y + 3)][x + (y + 3)] \\ &= a(x - y - 3)(x + y + 3) \end{aligned}$

c) $\begin{aligned} x^2y^2 - 4cxy - 2xy + 8c &= x^2y^2 - 2xy - 4cxy + 8c \\ &= xy(xy - 2) - 4c(xy - 2) \\ &= (xy - 2)(xy - 4c) \end{aligned}$

One of the uses of factoring is in finding the lowest common multiple of a group of algebraic expressions. The **lowest common multiple (L.C.M.)** is defined as the expression which has the smallest number of factors contained in each of the given expressions. To find the L.C.M. we simply factor each expression and collect each factor the largest number of times it occurs in any one expression.

Illustration. Find the L.C.M. of $(3x - 6)$, $3x^2 - 12x + 12$, and $x^3 - 8$.
Solution.

$$\begin{aligned} 3x - 6 &= 3(x - 2) \\ 3x^2 - 12x + 12 &= 3(x - 2)(x - 2) \\ \underline{x^3 - 8 = (x - 2)(x^2 + 2x + 4)} \\ \text{L.C.M.} = 3(x - 2)(x - 2)(x^2 + 2x + 4) \end{aligned}$$

The factors have been written in "columns" in order to ease the determination of the number of times each factor must be used in the L.C.M.

9. Fractions. An algebraic fraction is the indicated quotient of two algebraic expressions. The simplest fraction is a/b in which a is called the **numerator** and b is called the **denominator**. The fraction $1/x$ is known as the **reciprocal** of x. In all our discussion of fractions, the *denominator cannot be zero*. If for some values of the letters the denominator becomes zero, we say that the fraction is not defined for those values.

We shall start with the following properties of fractions:

1. *Multiplying, or dividing, both the numerator and denominator of a fraction by the same number or expression does not change the value of the fraction.*

2. *Changing the signs of an* **even** *number of* **factors** *does* **not** *change the sign of the fraction, whereas changing the sign of an* **odd** *number of* **factors changes** *the sign of the fraction.*

Examples:

a) $\dfrac{a + b}{a - b} = \dfrac{a + b}{a - b} \cdot \dfrac{a + b}{a + b} = \dfrac{(a + b)^2}{a^2 - b^2}$

b) $\dfrac{1}{x - y} = \dfrac{-1}{-(x - y)} = -\dfrac{1}{-(x - y)} = -\dfrac{1}{y - x} = \dfrac{-1}{y - x}$

c) $\dfrac{(x - y)(a - b)}{(y - x)(b + a)} = -\dfrac{(x - y)(a - b)}{(x - y)(b + a)} = \dfrac{b - a}{b + a}$

In considering the subject of algebraic fractions we are concerned with the four fundamental operations and the reduction of the fractions to their lowest terms. **Addition** and **subtraction** are accomplished by reducing the fractions to the lowest common denominator (L.C.D.) and then finding the algebraic sum of the resulting numerators:

$$\frac{a}{d} + \frac{b}{d} = \frac{a + b}{d} \quad \text{and} \quad \frac{a}{d} - \frac{b}{d} = \frac{a - b}{d}$$

Finding the L.C.D. of a set of fractions is easily accomplished by finding the L.C.M. of the denominators involved.

Examples:

a) $\dfrac{3}{x - 2} - \dfrac{4x + 1}{x^2 - 4} + \dfrac{5}{x + 2}$

L.C.D. $= (x - 2)(x + 2)$; then

$$\frac{3(x + 2)}{(x - 2)(x + 2)} - \frac{4x + 1}{(x - 2)(x + 2)} + \frac{5(x - 2)}{(x - 2)(x + 2)}$$

$$= \frac{3x + 6 - 4x - 1 + 5x - 10}{(x - 2)(x + 2)}$$

$$= \frac{4x - 5}{(x - 2)(x + 2)}$$

b) $\dfrac{x-2}{x^2-x-6} + \dfrac{x+1}{x^2+x-2} + \dfrac{x+3}{x^2-4x+3}$

L.C.D. $= (x-3)(x+2)(x-1)$; then

$$\dfrac{(x-2)(x-1) + (x+1)(x-3) + (x+3)(x+2)}{(x-3)(x+2)(x-1)}$$

$$= \dfrac{x^2 - 3x + 2 + x^2 - 2x - 3 + x^2 + 5x + 6}{(x-3)(x+2)(x-1)}$$

$$= \dfrac{3x^2 + 5}{(x-3)(x+2)(x-1)}$$

The **product** of two fractions is the fraction whose numerator is the product of their numerators and whose denominator is the product of their denominators.

$$\frac{a}{b} \cdot \frac{x}{y} = \frac{ax}{by}$$

To obtain the **quotient** of two fractions, invert the divisor and multiply.

$$\frac{a}{b} \div \frac{x}{y} = \frac{a}{b} \cdot \frac{y}{x} = \frac{ay}{bx}$$

Examples:

a) $\dfrac{a^2-b^2}{4a} \cdot \dfrac{8a}{a^2+2ab+b^2} = \dfrac{(a-b)(a+b)}{4a} \cdot \dfrac{\overset{2}{8a}}{(a+b)(a+b)}$

$$= \dfrac{2(a-b)}{a+b}$$

$$\left[\dfrac{x-y}{16c} \div \dfrac{x^3+y^3}{8c^2}\right] \cdot \dfrac{x^2-xy+y^2}{c}$$

$$= \dfrac{x-y}{16c} \cdot \dfrac{8c^2}{(x+y)(x^2-xy+y^2)} \cdot \dfrac{x^2-xy+y^2}{c}$$

$$= \dfrac{x-y}{2(x+y)}$$

10. Simplification of Fractions. A fraction may become very complex if the numerator or denominator, or both, are fractions. In such cases it is desirable to reduce them to their simplest form. In these complex fractions the place of the major division must be made clear;

this is usually done by a heavy rule. The simplification is accomplished by applying the above principles in a systematic order, as in the following illustrations.

Illustrations.

1) Simplify $\dfrac{\dfrac{1-c}{c} + \dfrac{c}{1+c}}{\dfrac{c}{c+1} + \dfrac{c-1}{c}}$.

Solution.

$$\frac{\dfrac{1-c}{c} + \dfrac{c}{1+c}}{\dfrac{c}{c+1} + \dfrac{c-1}{c}} = \frac{\dfrac{(1-c)(1+c)+c^2}{c(1+c)}}{\dfrac{c^2+(c-1)(c+1)}{c(c+1)}}$$

$$= \frac{1-c^2+c^2}{c(1+c)} \cdot \frac{c(c+1)}{c^2+c^2-1}$$

$$= \frac{1}{2c^2-1}$$

2) Simplify $1 - \dfrac{1}{2 - \dfrac{1}{3 - \frac{1}{4}}} = N$.

Solution.

$$N = 1 - \frac{1}{2 - \dfrac{1}{\dfrac{12-1}{4}}}$$

$$= 1 - \frac{1}{2 - \dfrac{4}{11}} = 1 - \frac{1}{\dfrac{22-4}{11}}$$

$$= 1 - \frac{11}{18} = \frac{18-11}{18} = \frac{7}{18}$$

11. Operations on Radicals. If radicals have the same index and the same radicand, they are called **like radicals** and may be added or subtracted as may other like expressions. If radicals are **unlike,** their algebraic sum can only be indicated. In order to perform the algebraic addition, we first reduce the radicals to their simplest form.

Illustrations.
1) Find the sum of $2\sqrt{x} + \sqrt{4x} - \sqrt{36x^3}$.
Solution.
$$2\sqrt{x} + \sqrt{4x} - \sqrt{36x^3} = 2\sqrt{x} + 2\sqrt{x} - 6x\sqrt{x}$$
$$= (4 - 6x)\sqrt{x}$$

2) Find the sum of $\sqrt{ax^2} - x\sqrt{4a} + 3a\sqrt{25ax^2}$

Solution.

$$\sqrt{ax^2} - x\sqrt{4a} + 3a\sqrt{25ax^2} = x\sqrt{a} - 2x\sqrt{a} + 15ax\sqrt{a}$$
$$= (15ax - x)\sqrt{a}$$

The **product** of two radicals is obtained by applying the laws discussed in Section 6. Let us now illustrate multiplication involving radicals.

Illustrations.

1) Find the product $(\sqrt{15})(\sqrt{\frac{5}{3}})$.

Solution.

$$(\sqrt{15})(\sqrt{\tfrac{5}{3}}) = \sqrt{15(\tfrac{5}{3})} = \sqrt{25} = 5$$

2) Find the product $(\sqrt[3]{18})(\sqrt{6})$.

Solution.

$$(\sqrt[3]{18})(\sqrt{6}) = (18^{\frac{1}{3}})(6^{\frac{1}{2}}) = (18)^{\frac{2}{6}}(6)^{\frac{3}{6}}$$
$$= \sqrt[6]{18^2 \cdot 6^3} = \sqrt[6]{3^4 \cdot 2^2 \cdot 3^3 \cdot 2^3}$$
$$= \sqrt[6]{3^7 \cdot 2^5} = 3\sqrt[6]{3 \cdot 32}$$
$$= 3\sqrt[6]{96}$$

3) Find the product $(2\sqrt{x} - 3\sqrt{y})(7\sqrt{x} + 6\sqrt{y})$.

Solution.

$$
\begin{array}{r}
2\sqrt{x} - 3\sqrt{y} \\
7\sqrt{x} + 6\sqrt{y} \\
\hline
14x - 21\sqrt{xy} \\
+ 12\sqrt{xy} - 18y \\
\hline
14x - 9\sqrt{xy} - 18y
\end{array}
$$

The **division** of radicals usually includes the operation called **rationalizing the denominator,** that is, removing all radicals from the denominator. This is accomplished by multiplying both numerator and denominator by a rationalizing factor. For this purpose we recall the very useful product

$$(a + b)(a - b) = a^2 - b^2$$

Illustrations.

1) Simplify $\sqrt{3} \div 7\sqrt{2}$.

Solution.

$$\frac{\sqrt{3}}{7\sqrt{2}} = \frac{\sqrt{3}}{7\sqrt{2}} \cdot \frac{\sqrt{2}}{\sqrt{2}} = \frac{\sqrt{6}}{7 \cdot 2} = \frac{1}{14}\sqrt{6}$$

2) Simplify $(\sqrt{5} + x) \div (\sqrt{5} - x)$.

Solution.

$$\frac{\sqrt{5} + x}{\sqrt{5} - x} = \frac{\sqrt{5} + x}{\sqrt{5} - x} \cdot \frac{\sqrt{5} + x}{\sqrt{5} + x} = \frac{(\sqrt{5} + x)^2}{5 - x^2}$$

$$= \frac{5 + x^2 + 2x\sqrt{5}}{5 - x^2}$$

3) Simplify $(\sqrt[6]{c} \div \sqrt[4]{c^2}) \cdot \sqrt{c^2}$.

Solution.

$$(\sqrt[6]{c} \div \sqrt[4]{c^2})(\sqrt{c^2}) = c^{\frac{1}{6}} \cdot c^{-\frac{2}{4}} \cdot c^{\frac{2}{2}}$$
$$= c^{\frac{1}{6} - \frac{1}{2} + 1} = c^{\frac{4}{6}} = c^{\frac{2}{3}}$$
$$= \sqrt[3]{c^2}$$

12. Imaginary Numbers. We have seen that the cube root of -8 has a meaning and a value, -2. But what about the square root of -4? Is there a number such that when multiplied by itself the result is -4? In the real-number system there is not; neither $+2$ nor -2 will yield the desired result. The operation of extracting an *even root of a negative number* requires us to define a new number which is called an **imaginary number.** It is a number which has the property that when multiplied by itself it yields a negative number. We let *i* denote the **imaginary unit** $\sqrt{-1}$; then we have

$$i^2 = (\sqrt{-1})^2 = -1$$

This permits us to write the square root of a negative number as the product of a real number and the imaginary number *i*, thus:

$$\pm\sqrt{-4} = \pm i\sqrt{4} = \pm 2i$$

Suppose we raise the number *i* to successive positive integral powers; we obtain

$$
\begin{array}{ll}
i = \sqrt{-1} & i^5 = i \\
i^2 = -1 & i^6 = -1 \\
i^3 = -i & i^7 = -i \\
i^4 = 1 & i^8 = 1
\end{array}
$$

Thus we see that the integral powers of *i* can be reduced to i, -1, $-i$, or 1.

The combination of an imaginary number and a real number of the form **a + bi** is called a **complex number.** We shall discuss complex numbers in Chapter III. At the present we are interested in simplifying expressions containing the imaginary quantity.

Illustrations.
1) Simplify $i^5 + 3i^4 - 5i^3 + 6i^2 - 3i$.
Solution.

$$i^5 + 3i^4 - 5i^3 + 6i^2 - 3i = i + 3 + 5i - 6 - 3i$$
$$= 3i - 3$$

2) Simplify $\dfrac{2 - \sqrt{-3}}{2 + \sqrt{-3}}$.

Solution.

$$\frac{2 - \sqrt{-3}}{2 + \sqrt{-3}} = \frac{2 - i\sqrt{3}}{2 + i\sqrt{3}} \cdot \frac{2 - i\sqrt{3}}{2 - i\sqrt{3}}$$

$$= \frac{4 - 4i\sqrt{3} + 3i^2}{4 - 3i^2}$$

$$= \frac{4 - 4i\sqrt{3} - 3}{4 + 3} = \frac{1 - 4i\sqrt{3}}{7}$$

13. Comparison. A fundamental exercise of algebra is the comparison of algebraic expressions. The reader should already be familiar with the symbols $=$, which is read "is equal to," and $\neq$, which is read "is not equal to." Any statement of equality is called an *equation*. The expressions are called *members* or *sides* of the equality and we speak of them as the left member or right member and as the left-hand side or right-hand side. There are five axioms which are assumed for equality.

$\mathbf{E_1}$. *Reflexive* property: $a = a$.
$\mathbf{E_2}$. *Symmetric* property: If $a = b$, then $b = a$.
$\mathbf{E_3}$. *Transitivity* property: If $a = b$ and $b = c$, then $a = c$.
$\mathbf{E_4}$. *Addition* property: If $a = b$ and $c = d$, then $a + c = b + d$.
$\mathbf{E_5}$. *Multiplication* property: If $a = b$ and $c = d$, then $a \cdot c = b \cdot d$.

In making statements of the kind made in the above axioms we shall assume that the letters represent *arbitrary* quantities* which permit complete freedom in substituting for the letters, and in particular that any quantity may be substituted for an equal quantity.

We call a simple comparison of either $a = b$ or $a \neq b$ a **dichotomy.** We shall also make a more extensive comparison of two quantities. A **trichotomy** is a comparison of two quantities in which exactly one of the following three relations holds:

---

*We use the word "quantity" in the sense of the dictionary definition: "an indefinite amount or number"; "the subject of a mathematical operation."

O_1. $a = b$, $a < b$, $b < a$.

For the present we shall assume two axioms for trichotomy.

O_2. *Transitivity* property: If $a < b$ and $b < c$, then $a < c$.

O_3. *Addition* property: If $a < b$, then $a + c < b + c$.

Examples:
a) Since $2 < 3$ and $3 < 3.5$, then $2 < 3.5$.
b) Since $3 < 4$, then $3 + 5 < 4 + 5$ or $8 < 9$.

The trichotomy property permits us to order quantities according to size or according to a manner in which they were selected. Thus, for example, we say that 2 is less than 3, $(2 < 3)$, or that a precedes b, $(a < b)$, if we have selected a to be first and b to be second. We shall also make use of the "greater than" sign and note that

$$a < b \quad \text{and} \quad b > a$$

are two expressions which state the same thing.

14. Mistakes.   The following is a list of common mistakes made by students when working with the subject matter of this chapter. We shall indicate that two quantities are not equal ($\neq$). Students, however, make the error of indicating that the quantities are equal. It is suggested that the student read this list, understand why the $\neq$ is used, and correct the errors.

1. $|-3| \neq -3$
2. $3^2 \cdot 3^3 \neq 9^5$
3. $a^2 \cdot b^5 \neq (ab)^7$
4. $x + y - 3(z + w) \neq x + y - 3z + w$
5. $a + (x - 3y) \neq a + x + 3y$
6. $3a + 4b \neq 7ab$
7. $7 - (-3) \neq 7 - 3$
8. $3x^{-1} \neq \dfrac{1}{3x}$
9. $\sqrt{x^2 + y^2} \neq x + y$
10. $\dfrac{x + y}{x + z} \neq \dfrac{y}{z}$
11. $\dfrac{1}{x - y} \neq -\dfrac{1}{x + y}$
12. $\dfrac{x}{y} + \dfrac{r}{s} \neq \dfrac{x + r}{y + s}$

13. $x\left(\dfrac{a}{b}\right) \neq \dfrac{ax}{bx}$

14. $\dfrac{xa + xb}{x + xd} \div x \neq \dfrac{a + b}{1 + d}$

15. $\sqrt{-x}\sqrt{-y} \neq \sqrt{xy}$

Summary of Symbols	
=, is equal to	$|n|$, absolute value of n
≠, is not equal to	$\sqrt{a}$, square root of a
<, is less than	a^n, n^{th} power of a
>, is greater than	$\sqrt[n]{a}$, n^{th} root of a
≤, is less than or equal to	i, imaginary unit
≥, is greater than or equal to	±, plus or minus

+, −, ×, ÷, arithmetic operations
( ), [ ], { }, signs of grouping

15. Exercise I

1. Simplify by removing the signs of grouping:

   a) $(3 + 5) - (6 - 2) - (-3 + 1)$

   b) $-\dfrac{1}{2} - \left\{\dfrac{3}{4} - 3\left[\dfrac{1}{3} - \left(\dfrac{3}{4} - \dfrac{7}{6}\right)\right]\right\}$

   c) $x - \{2x - [x - (2 - x) + 1] - 1\}$

   d) $3(x - 2) - 2(-x - 1) - 5(2x - 1) + 5x$

   e) $a\{a[a(a - 1)]\}$

2. Find the numerical value:

   a) $[x(x - y) + y(x + y)] - [x(x + y) - y(x - y)]$
      if $x = 1, y = 3$

   b) $\dfrac{3xy^2 - 3x^2y}{2(5 - 3x)}$ if $x = 2, y = 1$

   c) $|x| - |x + 2| + |-x| - 2|1 - x|$ if $x = -1$

   d) Problem 1(e) if $a = 1$

   e) $5x - \{2x - [2x - (x + 2x) - (2x - [x + 1])]\}$ if $x = 1$

3. Perform the indicated operations:

   a) $(3ax^2y)(-2a^2xy)$ b) $(-2ab^2)^3$

   c) $\dfrac{48a^2(x^3y)^4}{16(ax)^2y^2}$ d) $\left(\dfrac{x^{2n-1}}{y^{m+1}}\right)^3 \div \left(\dfrac{x^{n-1}}{y^m}\right)^2$

   e) $\sqrt[4]{x^2 - 2xy + y^2}$ f) $\sqrt{2} - \sqrt{8} + \sqrt{128}$

4. Multiply:
   a) $x^2 - 3xy - 2y^2$ by $x^2 - xy + 3y^2$
   b) $2x^2 - xy + y$ by $y - xy + x$
   c) $x^3 - 3x^2y + 3xy^2 - y^3$ by $x - y$
   d) $3 - i$ by $2 + i$ where $i^2 = -1$
   e) $(3.2x - 1.4y)$ by $(1.4x - 3.2y)$

5. Factor the following expressions:
   a) $x^2 + xy + xz + yz$ b) $x^2 - y^2 + (x + y)^2$
   c) $y^2 + y - 12$ d) $a^3 - 27b^3$
   e) $x^2 + 6x - 91$ f) $x^2 - 7x + 10$
   g) $3x^2 - 10xy + 3y^2$ h) $6x^2 + 19xy + 10y^2$
   i) $8a^3b^3 - 36a^2b^2x + 54abx^2 - 27x^3$
   j) $81ax^3 - 27bx^3 + 8by^3 - 24ay^3$

6. Divide:
   a) $x - 1$ by $x^{\frac{1}{2}} - 1$
   b) $x^3y^0 - 27x^0y^3$ by $x - 3y$
   c) $4x^4 - 19x^3y + 2x^2y^2 + xy^3 - 6y^4$ by
         $-4xy - 3y^2 + x^2$

7. Combine into one fraction and simplify:
   a) $\dfrac{3x - y}{x^2 - y^2} - \dfrac{3}{x + y}$ b) $\dfrac{2}{a - b} - \dfrac{2}{a + b}$

   c) $\dfrac{x + 1}{x^2 - x - 6} - \dfrac{x - 4}{x^2 - 4x + 3} + \dfrac{x + 4}{x^2 + x - 2}$

   d) $\dfrac{x + y}{6x^2 + 19xy + 10y^2} - \dfrac{2x - 5y}{3x^2 - xy - 2y^2} - \dfrac{3x - 2y}{2x^2 + 3xy - 5y^2}$

   e) $\left(\dfrac{x^2}{a^2} + 2\dfrac{x}{a} + 1\right) \div \left(\dfrac{x^2}{a^2} - 2\dfrac{x}{a} + 1\right)$

   f) $\dfrac{\dfrac{a}{b} + \dfrac{x}{y}}{\dfrac{a}{b} - \dfrac{x}{y}}$ g) $\dfrac{1}{2} - \dfrac{3}{4}$

   g) $\dfrac{\dfrac{1}{2} - \dfrac{3}{4}}{\dfrac{2}{5} - \dfrac{2}{3}}$

   h) $\dfrac{x}{1 - \dfrac{1}{1 - x}} - \dfrac{x^2}{x - \dfrac{x}{1 - x}}$

8. Multiply:
   a) $\sqrt{a} + \sqrt{b}$ by $\sqrt{a} - \sqrt{b}$
   b) $\sqrt{x} + 3\sqrt{y}$ by $\sqrt{3} - 3\sqrt{y}$
   c) $\sqrt{2x} - 2\sqrt{x}$ by $\sqrt{2} + 2\sqrt{x}$

9. Rationalize the denominator:

    a) $\dfrac{3}{\sqrt{2}}$ *b)* $\dfrac{\sqrt[3]{2}}{\sqrt{8}}$

    c) $\dfrac{\sqrt{3}+1}{\sqrt{2}-\sqrt{5}}$ *d)* $\dfrac{\sqrt{x}+\sqrt{y}}{\sqrt{x}-\sqrt{y}}$

    e) $\dfrac{\sqrt{a+b}-5\sqrt{a-b}}{3\sqrt{a+b}-2\sqrt{a-b}}$ *f)* $\dfrac{3+2\sqrt{x}}{2+3\sqrt{x}}$

10. Simplify:

    a) $\sqrt{-1}\,\sqrt{-2}\,\sqrt{-3}\,\sqrt{-4}$ b) $\dfrac{2-\sqrt{-2}}{2+\sqrt{-2}}$

    c) $(3+i)(2-i^3)$ d) $(2+i)\div(\sqrt{2}-i)$

II

BASIC CONCEPTS OF SETS

16. Introduction. Mathematics has frequently been referred to as the queen of the sciences. Although mathematics is a tool for science and much of the application of mathematics is tied to the physical world, most of mathematics as such is very abstract. In many respects mathematics is a language based on a logical structure. In our study of the subject we introduce technical terms, define words to build our vocabulary, and form sentences, much the same as when we study the English language. In order to build a logical structure and a mathematical language, however, we must first accept a small number of words to be undefined so that we may define others in terms of these. The reader may already have encountered such undefined words, e.g., *point* and *line* in geometry. We shall, of course, form mental pictures of these undefined words and give examples, but the words will nevertheless remain undefined. In our discussions we shall also use the normal words of the English language and accept their usual meanings. As we build our language we shall also make extensive use of symbols, which will become a convenient form of shorthand.

17. Statements. In mathematics we make statements which are usually assertions of a truth. As an example we say, "The number four is an even number." We implicitly assert that this statement is true. Perhaps we should say, "It is true that four is an even number." This is awkward, however, as is, "It is false that four is an odd number," when we could simply say, "The number four is not an odd number."

Some of the statements we make will be assumed to be true; these statements are called **axioms.** Other statements will be proved; these are called **theorems.** One of the most frequently occurring statements in mathematics is of the form "If ..., then ...." We shall discuss this form at length in Chapter X.

Examples:
a) If $x = 2$, then $x^2 = 4$.
b) If a is divisible by 2, then $3a$ is divisible by 2.
c) If x is an even integer, then $3x$ is not an odd integer.

Statements of this form are called **implications.** The usual mathematical notation is to let p represent the statement that follows *if* and to let q represent the statement that follows *then.* The standard form is

$$\text{"If } p, \text{ then } q.\text{"}$$

Illustration. Determine p and q in the implication, "If two triangles are congruent, then they are similar."
Solution. In the standard form, "If p, then q," p here stands for "two triangles are congruent" and q stands for "they are similar."

The interchange of p and q in the standard form results in a converse statement, which we shall give a formal definition:

Definition. *The implication* "If q, then p" *is the* **converse** *of the implication* "If p, then q."

Example:
Implication: If two triangles have the three sides of one equal respectively to the three sides of the other, then the triangles are congruent.
Converse: If two triangles are congruent, then the three sides of one are equal respectively to the three sides of the other.

Although the implication may be true, the converse is not necessarily true.

Example:
Implication: If two triangles are congruent, then they are similar.
Converse: If two triangles are similar, then they are congruent.

Let us consider the following two statements:

The number 2 is even;
The number 2 is not even.

The relationship between them is that one is a "not" statement of the other. If we are given a statement p, then the statement *not-p* is called the **negation.** In the statements above we see that both cannot be true. We shall therefore give the following meaning to negation.

Definition. *The* **negation** *of a given statement is a statement such that (a) if the given statement is true, then the negation is false; (b) if the given statement is false, then the negation is true.*

Negation plays an important role in implications and in logic in general. By using negation we can now form statements concerning *not-p* and *not-q*. Suppose we start with an implication, "If p, then q," next form the statements *not-p* and *not-q*, and then interchange the roles of the new statements to form the new implication, "If *not-q*, then *not-p*." This statement will be given a special name.

Definition.   *The implication* "If *not-q*, then *not-p*" *is the* **contrapositive** *of the implication* "If p, then q."

Examples:
a) *Implication:* If $x = 5$, then $x^2 = 25$.
   Contrapositive: If $x^2 \neq 25$, then $x \neq 5$.
b) *Implication:* If $x + 3 = 5$, then $x = 2$.
   Contrapositive: If $x \neq 2$, then $x + 3 \neq 5$.
We can now state a basic law of logic.
   An implication and its contrapositive are either both true or both false.

You will notice the difference between a contrapositive in relation to an implication and a converse in relation to an implication. Be sure that you can state this difference.

If the implication and its converse are both true, then p and q are **equivalent** and we frequently use a stronger expression to state the implication. This expression uses the words "if and only if" and we write "q if and only if p" or "p if and only if q."
Examples:
a) $x + 3 = 5$ if and only if $x = 2$.
b) Two triangles are similar if and only if they are mutually equiangular.
c) An integer is even if and only if it is divisible by 2.

Illustration.   Write Example (b) above as an "if p, then q" statement and write the converse.
Solution.   The implication is
   If two triangles are mutually equiangular, then they are similar.
The converse is
   If two triangles are similar, then they are mutually equiangular.

18. The Concept of a Set.   The fundamental concepts of modern mathematics are based on the idea of a **set**. We shall therefore consider the idea of a set to be so basic that it is not defined. However, in order to form a mental picture, let us think of a set as a collection of objects which are described in some specified manner.

Consider the following examples:

a) The set of all boys under 10 years of age,
b) The set of all the states in the United States,
c) The set of positive integers,
d) The set of all integers divisible by 2,
e) The set of all the people in this room,
f) The set of positive integers less than 5,
g) The set of letters in the English alphabet,
h) The set of all points that lie on a given line,
i) The set of fractions between 1 and 2,
j) The set consisting of one dog and one boy.

Note that in each of the samples there is a descriptive definition which provides a means for determining whether or not an object belongs to the given set. Thus, in Example (a), a 7-year-old boy would belong to the set but a 12-year-old boy or a 6-year-old girl would not. If an object belongs to a given set, it is called an **element** of this set. We use the notation

$$a \in A$$

to mean "a is an element of the set A." This notation can also be read "a belongs to the set A" and "a is contained in the set A." The expression

$$a \notin A$$

is used to indicate that a is not an element of the set A.

There are two common methods used to denote a set. The first is simply to list all the elements of the set, enclosing them in braces; e.g.,

$$A = \{1, 3, 5, 7, 9\}$$

means the set of all positive odd integers less than 10. The second method is to enclose the defining statement in braces; e.g.,

$$A = \{x \text{ such that } x \text{ is a positive odd integer less than } 10\}$$

where x is used as a symbol for an arbitrary element of the set. This notation can be shortened by letting a vertical bar, |, represent the words "such that." The above example can then be written

$$A = \{x | x \text{ is a positive odd integer} < 10\}$$

19. Open Sentences—Equations.   We shall now apply the concept of sets to a fundamental topic of algebra.

Definitions. *A* **variable** *is a letter used to represent an arbitrary element of a set which contains more than one element. If a set contains only one element, then the letter representing this element is called a* **constant.**

Examples:
a) If $X = \{2, 3, 4\}$, then $x \in X$ is a variable which stands for 2, 3, or 4.
b) If $A = \{\pi\}$, then $a \in A$ is a constant; namely, $a = \pi$.

Definitions.* *The set whose elements may serve as replacements for a variable is called the* **domain** *or* **replacement set** *of the variable. Members of the domain are called the* **values** *of the variable.*

We pointed out in Section 17 that mathematics concerns itself with making statements. These statements are in the form of sentences. An important class of sentences are those which contain variables, and these are given a special name.

Definition. *Sentences containing variables are called* **open sentences.**

Example: $x + 5 = 9$
Since the open sentence contains a variable it entertains further discussion. The usual question is, "For what values of the variable is the sentence a true statement?" To answer this question we need to specify the domain of the variable, and the question should include a definition of the replacement set.

Definition. *Elements of the domain (replacement set) for which the open sentence is true are called the* **solution set** *or* **truth set** *of the open sentence over that domain.*

Each element of the solution set is said to *satisfy* the open sentence. It is also called a *root* or *solution* of the open sentence.
Example: If $S = \{1, 2, 3, 4, 5, 6, 7, 8\}$ is the domain of the variable x, the open sentence "x belongs to S and $x + 3 = 5$" is satisfied by the element 2 of S and $\{2\}$ is the solution set or truth set. $x = 2$ is a solution of the sentence.

When the open sentence is a statement of equality it is an equation, and if the equality is true for only certain values of the variable over its domain it is called a *conditional equality*. If the open sentence is

---

*Some authors also refer to the domain of a variable as the *universe* of the variable.

true for all values of the variable over its domain, it is called an *identity*.

Example. If $X = \{x \mid x$ is an integer$\}$ is the domain of the variable x, then the open sentence, "x belongs to X and $x^2 - 4 = (x - 2)(x + 2)$" is true for all values of x; the solution set is X and the open sentence is an identity.

20. Operations for Equations. The form of an equation may be changed by an algebraic operation. However, we must be careful not to destroy its original meaning. If we change an equation to another form, we then have a *derived* equation.

> **Definition.** *A derived equation is* **equivalent** *to the original equation if it contains all the roots of that equation and no more.*

With this definition it should be clear that equivalent equations have the same truth set.

There are two main operations which lead to equivalent equations:

1. *We may add the same number or expression to, or subtract the same number or expression from, both members of the equation.*

2. *We may multiply or divide both members of the equation by the same number or expression, providing it is not zero or does not contain the variable.*

Examples:

a) $2x = 5$　and　$2x + 3 = 8$　are equivalent.

b) $2x = 5$　and　$4x = 10$　are equivalent.

The first of these operations leads to the well-known concept of **transposing** terms from one side to the other, since this is simply the operation of subtracting the same quantity from both sides.

Example: If we transpose 2 from the left member to the right member in the equation $3x + 2 = 5$ we obtain $3x = 5 - 2 = 3$, which can also be obtained by subtracting 2 from both sides.

Although the multiplication of both members by an expression containing the variable does not lead to equivalent equations, it is sometimes done and may result in an equation which has roots that are *not* roots of the original equation. Such roots are called **extraneous roots.** The derived equation is said to be **redundant** with respect to the original equation if it contains all the roots of that equation plus some others.

We may also divide both members of an equation by an expression which contains the variable, but in so doing we may lose some roots. The derived equation is said to be **defective** with respect to the

original equation if it does not have all the roots of that equation.
Examples:

a) Consider the equation $x + 2 = 3$, which has the root $x = 1$. multiply both members by $x + 1$; the result is $x^2 + 3x + 2 = 3x + 3$ or $x^2 = 1$, which has the roots $x = \pm 1$. The root -1, however, is not a root of the original equation.

b) Consider the equation $(x - 2)(x + 2) = 0$, which has roots $x = 2$ and $x = -2$. Divide both members by $x - 2$ to get $x + 2 = 0$, which has the root $x = -2$ but does not have the other root.

One of the first equations we meet is the **polynomial equation** in one variable:

$$a_0x^n + a_1x^{n-1} + a_2x^{n-2} + \ldots + a_{n-1}x + a_n = 0$$

where n is a positive integer and the coefficients $a_0, a_1, a_2, \ldots, a_n$ are constants. The **degree** of this equation is the degree of the term that is highest. We give special names to some of the lower-degree equations:

Degree	Name	Equation
first	linear	$ax + b = 0$
second	quadratic	$ax^2 + bx + c = 0$
third	cubic	$ax^3 + bx^2 + cx + d = 0$
fourth	quartic	$ax^4 + bx^3 + cx^2 + dx + e = 0$

The operations are used to find the truth sets of open sentences.

Illustration. Solve for x in the equation $3x - 7 = 2x - 5$.
Solution.

Subtract $2x$ to obtain $x - 7 = -5$
Add 7 to obtain $x = 2$
Check: $3(2) - 7 = 2(2) - 5$ or $-1 = -1$

Procedures for finding truth sets of open sentences will be developed throughout the entire book.

21. Subsets. Consider the set of all positive integers,

$$A = \{1, 2, 3, 4, \ldots\}$$

A set of all the positive integers less than 7 will be a part of this set. Similarly, the set of all even positive integers will be a part of this set. This illustrates a very important concept of sets and we shall state this concept as a definition.

Definition. *Set* A *is a* **subset** *of set* B *if every element of* A *is an element of* B. *If* B *has elements that are not elements of* A, *then* A *is said to be a* **proper subset** *of* B.

We shall adopt a notation for this concept.*

$$A \subset B \text{ reads } \text{``}A \text{ is a subset of } B.\text{''}$$

The same notation can be read "*A* is contained in *B*" or "*A* is included in *B*."

Examples: Let $B = \{x | x \text{ is a positive integer}\}$.

a) If $A = \{x | x \text{ is a positive integer} < 5\}$, then $A \subset B$.

b) If $A = \{1, 2, 7\}$, then $A \subset B$.

c) If $A = \{x | x \text{ is a positive integer}\}$, then $A \subset B$.

Note that in the first two examples *A* is a proper subset of *B* but that in Example (c) *A* is not a proper subset of *B*.

Illustrations.

1) If $B = \{x | x \text{ is an integer}\}$, then is $A = \{x | x \text{ satisfies } 2x = 4\}$ a subset of *B*?

Solution. The set *A* consists of one element, which is the solution of the equation $2x = 4$. This solution is $x = 2$, which is an integer. Therefore, $A \subset B$.

2) If $B = \{x | x \text{ is an integer}\}$, then is $A = \{x | x \text{ satisfies } 2x = 5\}$ a subset of *B*?

Solution. The set *A* consists of the element which is the solution of the equation $2x = 5$. This solution is $x = 5/2$, which is *not* an integer. Therefore $A \not\subset B$.

Let us consider the set

$$A = \{x | x \text{ is a letter in the English alphabet}\}$$

and let V be the set of vowels,** $V = \{a, e, i, o, u\}$. Then we see that $V \subset A$. Let *C* be the set of consonants; then $C \subset A$. In this discussion the set *A* is a basic set which contains all the elements under discussion and deserves special recognition.

Definition. *The* **universal set** *in any discussion is the totality of members under consideration as elements of any set.*

---

*This is the generally accepted notation. Some authors make a distinction between "subset" and "proper subset" by using an underscore for subsets; i.e., subset: $A \subseteq B$; proper subset: $A \subset B$.

**We shall not consider *y* to be a vowel.

This universal set, which is usually denoted by U, will change from one discussion to another. In the above discussion it was the set of all the letters of the English alphabet; in another case it may be the set of integers, or perhaps the set of all the points in the plane.

If A is a proper subset of a universal set U, then there are elements in U which are not in A. In other words, if we take all the elements of A from U, there will be some elements left over; these left-over elements are given a special name.

Definition. *The* **complement** *of any set* A, *with respect to a universal set, is the set of elements of this universal set which do not belong to* A.

The complement of A is denoted by A' and is written

$$A' = \{x | x \notin A\}$$

Examples: Let $U = \{1, 2, 3, 4, 5, 6, 7\}$.
a) If $A = \{1, 2, 3, 4\}$, then $A' = \{5, 6, 7\}$.
b) If $A = \{1, 3, 5, 7\}$, then $A' = \{2, 4, 6\}$.
c) If $A = \{1, 2, 3, 4, 5, 6, 7\}$, then A' has no elements.

In the last example the complement of A has no elements. A set which has no elements is said to be the **empty set** or **null set** and is denoted by $\emptyset$. Since the null set has no elements we shall adopt the convention that it is a subset of all other sets.*

The definition of the complement of a set suggests the idea of finding the difference between two sets. We shall, however, need a formal definition for sets when the universal set is not involved.

Definition. *The* **difference** *of two sets A and B in the order* $A - B$ *is the set that contains those elements of A that do not belong to B:*

$$A - B = \{x | x \in A \text{ and } x \notin B\}$$

With this definition we see that $U - A = A'$ and $U - A' = A$.
Examples:
a) If $A = \{5, 6, 7, 8\}$ and $B = \{6, 7\}$, then $A - B = \{5, 8\}$.
b) $\{1, 2, 3\} - \{3, 4, 5\} = \{1, 2\}$
c) $\{3, 4, 5\} - \{1, 2, 3\} = \{4, 5\}$

We note from the examples that $A - B \neq B - A$ and that if $B \not\subset A$, there will be elements in B which have no bearing on $A - B$. We notice also that if $A \subset B$, then $A - B = \emptyset$ and as a special case

---

*Note that $\{0\} \neq \emptyset$ for the set $\{0\}$ has one element, the number zero, and therefore is not empty.

$A - U = \emptyset$. The following four properties follow directly from the definitions:

I. $A - \emptyset = A$ II. $A - A = \emptyset$
III. $\emptyset - A = \emptyset$ IV. $(A')' = A$

The examples we have been using indicate that sets can have varying numbers of elements; i.e., a set can have 3 elements, 6 elements, no elements, or any number of elements. In fact, a set can have an infinite number of elements, e.g., the set of all the points in a plane or the set of all the integers. If two sets have the same number of elements, they are said to be of the same size. We shall give this concept a more formal definition and at the same time introduce a very fundamental concept of mathematics.

Definition. *A* **one-to-one correspondence** *exists between two sets* A *and* B *if it is possible to associate each element of* A *with exactly one element of* B *and each element of* B *with exactly one element of* A.

Illustrations.
1) Establish the one-to-one correspondence between $A = \{1, 5, 7\}$ and $B = \{a, b, c\}$.
Solution. List the association

$$1 \leftrightarrow a, \; 5 \leftrightarrow b, \; 7 \leftrightarrow c$$

2) Is there a one-to-one correspondence between $A = \{2, 4\}$ and $B = \{r, s, t\}$?
Solution. We can associate the elements of *A* by $2 \rightarrow r$ and $4 \rightarrow s$. However, when we try to associate each element of *B* with exactly one element of *A*, there is one element of *B* for which there is no element of *A*. Therefore, the answer is no.

We can now give a precise definition concerning the size of two sets.

Definition. *Two sets* A *and* B *are the* **same size** *if there exists a one-to-one correspondence between their elements.*

Although two sets may be of the same size they are not necessarily equal; e.g., the set of 10 fingers and the set of 10 toes are the same size but we do not consider them to be equal. The concept of subsets gives us the basis for the determination of equality between sets.

Definition. *Two sets* A *and* B *are said to be* **equal** *if* $A \subset B$ *and* $B \subset A$.

It should be clear that if two sets are equal they are of the same size.

Furthermore, if one set has an element which is not in the other, then the two sets are not equal, $A \neq B$.

Definition.   *Two sets,* A *and* B, *are said to be* **disjoint** *if and only if* A *and* B *have no elements in common.*

Let us illustrate these definitions.

Examples: Consider the four sets

$$U = \{1, 2, 3, 4, 5, 6, 7, 8\} \qquad A = \{2, 4\}$$
$$B = \{1, 2, 3, 4\} \qquad C = \{5, 6, 7, 8\}$$

a) B and C have the same size but are not equal.

b) $A \subset B$

c) B and C are disjoint.

d) There is a one-to-one correspondence between the elements of B and those of C.

e) $C' = \{1, 2, 3, 4\} = B$

f) $B' = \{5, 6, 7, 8\} = C$

g) $A \subset C'$

Theorem.   *Let* A *be a set of elements of a universe* U. *Then* A $=$ U *if and only if* A$'$ *is the empty set.*

Proof.   Part I. If $A = U$, then $A \subset U$ and $U \subset A$ by definition. Since every element of U is also an element of A, then A' is empty.

   Part II. Since $A' = \emptyset$, then by definition $U \subset A, A \subset U$, and $A = U$.

22. Venn Diagrams.   A geometric interpretation of sets can be very helpful in visualizing sets and their properties. We shall therefore agree on a procedure for drawing diagrams which represent sets. Such diagrams are called Venn diagrams.* Let us consider each element of the universe U to be represented by a point and identify the set U with all the points within a rectangle. We shall represent

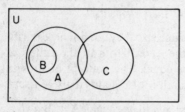

Fig. 1

any subset of U by the points inside a closed region, such as a circle, within the rectangle. In Figure 1 we show the universal set U and the three sets A, B, and C. The figure is drawn so that B is a subset of A, B and C are disjoint sets, and A and C have some elements in common.

---

*Named after the English logician John Venn (1834–1883).

In a Venn diagram the complement of a set is represented by the points in the rectangle outside of the circle which indicates the set.

23. Union and Intersection.   We shall now consider two important ideas concerned with operations on two or more sets, and as usual we shall begin with a definition.

> **Definition.**   *The* **union** *of two sets* A *and* B, *written* **A ∪ B,** *is the set of elements which belong to either* A *or* B *or to both* A *and* B.

Examples:

a) If $A = \{a, b, c\}$ and $B = \{d, e\}$, then $A \cup B = \{a, b, c, d, e\}$.

b) If $A = \{a, n\}$ and $B = \{n, c\}$, then $A \cup B = \{a, n, c\}$.

c) If $A = \{1, 3, 5, 7\}$ and $B = \{3, 5\}$, then $A \cup B = \{1, 3, 5, 7\}$.

Although the union of two sets may be thought of as the collection or sum of all the elements of the two sets, "sum" should not be understood in the mathematical sense; the common elements of the two sets are not counted twice in the union set. In example (b) above, the letter *n* occurred in both sets but in the union set we did not list *n* twice. Thus, the union set will contain all the elements of the two given sets but will list each element only once.

Since the union of two sets is a set, we can form another union of this set with a third set, $(A \cup B) \cup C$, and we can continue the idea to four or more sets.

> **Definition.**   *The* **intersection** *of two sets* A *and* B, *written* **A ∩ B** *is the set of elements which belong to both* A *and* B.

Examples:

a) $\{3, 5, 7\} \cap \{5, 6, 7, 8\} = \{5, 7\}$

b) $\{a, b, c\} \cap \{a, c, d\} = \{a, c\}$

c) $\{2, 4, 6\} \cap \{1, 3, 5\} = \emptyset$

d) $A \cap A' = \emptyset$

e) $A \cap A = A$

f) In Figure 2 the shaded area shows $C = A \cap B$.

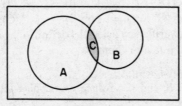

Fig. 2

The simple operations of forming the union and intersection of sets together with that of taking the difference, discussed in Section 21, obey some simple laws. Sets, subsets, and the laws of combination form an algebraic system which is usually referred to as Boolean algebra.* There is much more to Boolean algebra than what is presented in this chapter. However, let us list the basic laws and refer to them whenever we have the opportunity.

I. **Laws Concerning Union and Intersection:**

I.1. $A \cup A = A$ I.2. $A \cap A = A$

I.3. $A \cup U = U$ I.4. $A \cap U = A$

I.5. $A \cup \emptyset = A$ I.6. $A \cap \emptyset = \emptyset$

I.7. $A \cup B = B \cup A$ I.8. $A \cap B = B \cap A$

I.9. $A \cup (B \cup C) = (A \cup B) \cup C$ I.10. $A \cap (B \cap C) =$

I.11. $A \cup (B \cap C) =$ $(A \cap B) \cap C$

     $(A \cup B) \cap (A \cup C)$

I.12. $A \cap (B \cup C) = (A \cap B) \cup (A \cap C)$

(Note that the Commutative Laws are given by I.7 and I.8, the Associative Law by I.9 and I.10, and the Distributive Laws by I.11 and I.12.)

II. **Laws Concerning Complements:**

II.1. *The complement of the complement is the original set,* $(A')' = A.$

II.2. $A \cup A' = U$ II.3. $A \cap A' = \emptyset$ II.4. $U' = \emptyset$

II.5. *The complement of the union of two sets is the intersection of their complements,* $(A \cup B)' = A' \cap B'.$

II.6. *The complement of the intersection of two sets is the union of their complements,* $(A \cap B)' = A' \cup B'.$

III. **Laws Concerning Differences of Sets:**

III.1. $U - A = A'$ III.2. $A - U = \emptyset$

III.3. $A - \emptyset = A$ III.4. $\emptyset - A = \emptyset$

III.5. $A - A = \emptyset$ III.6. $A - B = A \cap B'$

III.7. $(A - B) - C = A - (B \cup C)$

III.8. $A - (B - C) = (A - B) \cup (A \cap C)$

III.9. $A \cup (B - C) = (A \cup B) - (C - A)$

III.10. $A \cap (B - C) = (A \cap B) - (A \cap C)$

The laws may be verified by using the definitions or Venn diagrams. We shall illustrate a few of them.

Illustration. Verify Law I.11.

Solution. Let A, B, and C be three overlapping sets, as shown in

---

*Named after the British mathematician George Boole (1815–1864).

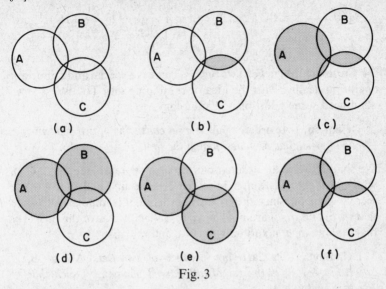

Fig. 3

Figure 3(a). Consider the left-hand side of the equality and first shade the intersection of B and C, as shown in Figure 3(b). Then shade $A \cup (B \cap C)$, as shown in Figure 3(c). Now consider the right-hand side. First shade $A \cup B$, Figure 3(d); then shade $A \cup C$, Figure 3(e); now form the intersection of these two shaded areas, Figure 3(f). Since Figures 3(c) and 3(f) are identical, the proposition is verified.

Illustration. Verify Law 6 of Differences of Sets.
Solution. $A - B = \{x \mid x \in A \text{ and } x \notin B\}$ by definition
$B' = \{x \mid x \notin B\}$ by definition
$A \cap B' = \{x \mid x \in A \text{ and } x \in B'\}$ by definition
$= \{x \mid x \in A \text{ and } x \notin B\}$ by Step 2
$= A - B$ by Step 1

Illustration. Verify Law 3 of Complements.
Solution. $A' = \{x \mid x \notin A\}$ by definition; consequently, A and A' have no elements in common and $A \cap A' = \emptyset$ by definition.

These laws can be used to derive other properties.

Illustration. Show that $(A \cap B) \cup (A \cap B') = A$.
Solution. $(A \cap B) \cup (A \cap B') = A \cap (B \cup B')$ by I.12.
$B \cup B' = U$ by II.2.
$A \cap U = A$ by I.4.

Illustration. Show that $A \cup (A' \cap B) = A \cup B$.

Solution. $A \cup (A' \cap B) = (A \cup A') \cap (A \cup B)$ by I.11.
$$= U \cap (A \cup B) \quad \text{by II.2.}$$
$$= A \cup B \quad \text{by I.4.}$$

24. Cartesian Product of Two Sets.   There is a special and important operation dealing with the idea of combining sets. To discuss it we shall need some additional terminology.

Definition.   *An* **ordered pair** *is a set consisting of two elements, one of which has been designated the first.*

We shall indicate an ordered pair by using parentheses thus, (x, y), which also specifies that x has been ordered first and that y is the second element. Since the pairs are ordered, it is important to note that (a, b) and (b, a) are not the same. In fact, we have the property that $(a, b) = (c, d)$ if and only if $a = c$ and $b = d$.

Definition.   *The* **Cartesian product** *of two sets* A *and* B, *written* A $\times$ B, *is the set of all ordered pairs* (x, y) *such that* x $\in$ A *and* y $\in$ B;

$$A \times B = \{(x, y) | x \in A, y \in B\}$$

The sign $\times$ is read "cross."
Examples:
a) If $A = \{a, b\}$ and $B = \{c, d\}$, then $A \times B = \{(a, c), (a, d), (b, c), (b, d)\}$.
b) If $A = \{1, 3\}$ and $B = \{2, 4\}$, then $A \times B = \{(1, 2), (1, 4), (3, 2), (3, 4)\}$.
c) If $A = \{1, 2\}$, then $A \times A = \{(1, 1), (1, 2), (2, 1), (2, 2)\}$.
Note that in Example (c) we have the two different elements $(1, 2)$ and $(2, 1)$.
Theorem.   *The Cartesian product of two different sets,* (A $\neq$ B), *is not commutative,* A $\times$ B $\neq$ B $\times$ A.
Proof.   Since $A \neq B$, there exists by definition at least one element in one of the sets which is not in the other. Let this element be $x_1 \in A$ and $x_1 \notin B$. The ordered pairs (x_1, y) where $y \in B$ would appear in $A \times B$ but would not appear in $B \times A$ since $x_1 \notin B$.
Example: Let $A = \{1, 2, 3\}$ and $B = \{2, 3\}$. Then

$$A \times B = \{(1, 2), (1, 3), (2, 2), (2, 3), (3, 2), (3, 3)\}$$
$$B \times A = \{(2, 1), (2, 2), (2, 3), (3, 1), (3, 2), (3, 3)\}$$

We see that $(1, 2)$ and $(1, 3)$, both elements of $A \times B$, are not elements of $B \times A$. Note that the two product sets are of the same size. Name the two elements of $B \times A$ that are not elements of $A \times B$.

The concepts presented in this chapter are very basic and we shall

use them throughout the book. The reader should have a firm understanding of the definitions and be very familiar with the symbols.

Summary of Symbols	
$a \in A$, a is an element of A	$\mid$, such that
$a \notin A$, a is not an element of A	$\{\ldots\}$, set
$B \subset A$, B is a subset of A	$\not\subset$, not a subset
A', complement of A	$A - B$, difference
$A \cup B$, union of A and B	$\emptyset$, null set
$A \cap B$, intersection of A and B	U, universal set
$A \times B$, Cartesian product of A and B	(x, y), ordered pair

25. Exercise II.

1. List the set defined by the following definitions:
   a) The set of positive integers smaller than 12.
   b) The set of even integers.
   c) A set of four boys whose first names begin with J.
   d) The set of prime numbers less than 20.
   e) The integers between 5 and 17.
   f) The vowels in the English alphabet.
   g) The solutions of the equation $x + 5 = 12$.
   h) The integer which satisfies the equation $2x = 3$.

2. Given the set A, find all the proper subsets $B \subset A$ under the following conditions:
   a) $A = \{1, 3, 5\}$ and B has two elements.
   b) $A = \{1, 2, 4, 5\}$ and the elements of B are even integers.
   c) $A = \{a, b, c\}$ and $B = \{x \mid x$ is a vowel$\}$.
   d) $A = \{3, 4, 8\}$ and $B = \{x \mid x$ is a prime number$\}$.
   e) $A = \{1, 3, 5\}$ and $B = \{x \mid x$ is an even integer$\}$.
   f) $A = \{$Jack, Joe, Jake$\}$ and the elements of B are boys whose first name ends in the letter e.

3. Show that there is a one-to-one correspondence between set A and set B if
   a) $A = \{1, 2, 3, \ldots, 26\}$ and $B = \{x \mid x$ is a letter in the English alphabet$\}$.
   b) $A = \{x \mid x$ is a positive integer$\}$ and $B = \{x \mid x$ is a negative integer$\}$.
   c) $A = \{x \mid x$ is a positive integer$\}$ and $B = \{x \mid x$ is a positive even integer$\}$.

4. Find the set resulting from the indicated operations:
   - a) $\{1, 2, 3\} \cup \{2, 3, 4\}$
   - b) $\{1, 2, 5\} \cup \{2, 5, 6\}$
   - c) $\{1, 3\} \cup \{2, 4\}$
   - d) $\{1, 2, 3\} \cap \{2, 3, 4\}$
   - e) $\{1, 2, 3, 4, 5\} \cap \{2, 4, 7\}$
   - f) $\{2, 4, 6, 8\} \cap \{1, 3, 5\}$
   - g) $\{2, 3, 4\} \cap \{1, 2, 3, 4\}$
   - h) $A \cup \emptyset \cap \emptyset$
   - i) $A \cup A'$
   - j) $A \cap A'$
   - k) $\{x | x \text{ is a positive integer}\} \cup \{x | x = 0, \text{ and } x \text{ is a negative integer}\}$

5. Let $B \subset A$. Find
   - a) $A \cup B$
   - b) $A \cap B$
   - c) $A' \cup B$
   - d) $A' \cap B$
   - e) $A \cup B'$
   - f) $A \cap B'$
   - g) $A' \cup B'$
   - h) $A' \cap B'$

6. Draw Venn diagrams for all parts of Problem 5.

7. Find $A \times B$:
   - a) $A = \{1, 2, 3\}, B = \{a, b\}$
   - b) $A = \{x, y, z\}, B = \{c, d\}$
   - c) $A = \{x_1, x_2\}, B = \{y_1, y_2\}$
   - d) $A = \{\text{Chevrolet, Ford}\}, B = \{\text{coupe, sedan, convertible}\}$
   - e) $A = B = \{\text{true, false}\}$

8. Use Venn diagrams to verify the following:
   - a) $A \cap B = B \cap A$
   - b) $(A')' = A$
   - c) $A \cap (B \cup B') = A$
   - d) $A \cap (B \cup C) = (A \cap B) \cup (A \cap C)$
   - e) If $A \subset B$ and $B \subset C$, then $A \subset C$
   - f) If $C' \subset B$, then $B' \subset C$

9. If $U = \{1, 2, 3, 4, 5, 6, 7, 8, 9\}$, find A' in the following cases:
   - a) $A = \{x | x \text{ is an even positive integer} < 10\}$
   - b) $A = \{x | 3 \leq x \leq 7\}$
   - c) $A = \{x | x - 3 = 5\}$
   - d) $A = \{1, 3, 5, 7, 9\}$
   - e) $A = \{x | x \text{ is larger than 4 but smaller than 8}\}$

10. Consider Figure 4, composed of a circle, trinagle, and rectangle

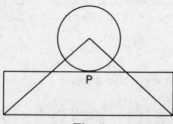

Fig. 4

placed with the circle tangent to the rectangle at P. Let A be
all the points inside and on the circle, B be all the points inside
and on the triangle, C be all the points inside and on the
rectangle, and let $U = A \cup B \cup C$. By shading the appropriate
area show that

a) $A' \cap C' < B$

b) $B' \cap C$ and $B' \cap A$ are disjoint.

c) $(A \cap B) \cap C = P$

11. Determine if the following statements are true or false:

a) $\{5, 7\} \cap \{1, 3, 5, 9\} = \{5\}$

b) $A \cup (B \cup C) = (A \cup B) \cup C$

c) If $B \subset A$, then $B \cup A = A$.

d) $(A \cap B)' = A' \cup B'$

e) $A \cap B = B \cap A$

f) If $3 \in A$, then $\{3\} \subset A$.

g) If $A = \{a\}$ and $B = \{b\}$, then $A \times B = \{(a, b)\}$.

h) If $A \subset B$ and $B \subset A$, then $A \neq B$.

i) Set A is a proper subset of B if and only if A is a subset of B.

j) If every element of A is an element of B, then $B \subset A$.

k) $(U \cup \emptyset)' = U$

l) $U - A' = A$

m) Sets $\{a, b\}$ and $\{b, c\}$ are disjoint.

n) The converse of "if p, then q" is "if q, then p".

o) $B \subset A$ if and only if $A' \subset B'$.

III

THE NUMBER SYSTEM

26. Introduction. The purpose of this chapter is to present the number system in a formal manner in order that we may better understand many procedures which have become intuitive. The number system was developed through a process of *extension*. Starting with the whole numbers (the positive integers) for the purpose of counting, it was discovered that the operation of subtraction could result in numbers that are not members of the set of whole numbers. In this way the number zero and the negative integers were added to the system. The operation of division made it necessary to establish the concept of a fraction, and then the process of taking a root called for some of the irrational numbers and eventually the complex numbers. The process of extension can also be related to the problem of finding the solutions of algebraic equations.

The number system has a very logical development and we shall present the first part of this development. No attempt will be made to cover the entire theory of numbers, which would require several books and a considerable knowledge of advanced mathematics.

27. The Fundamental Number Sets. We shall begin with the familiar numbers and discuss them using the concept of sets.

I.   The set of **natural numbers,** $N = \{1, 2, 3, 4 \ldots\}$

II. The set of **integers,** $I = \{\ldots, -3, -2, -1, 0, 1, 2, 3 \ldots\}$

III. The set of **rational numbers,** $Q = \{x | x = a/b,\ a \in I,\ b \in I,\ b \neq 0\}$

IV. The set of **real numbers,** $R = \{x | x$ can be expressed by a decimal expansion$\}$

V.   The set of **complex numbers,** $C = \{(a, b) | a \in R$ and $b \in R\}$

As we progress from I. to V. we see that each succeeding set contains more elements than its predecessors. In fact, from the definitions it should be clear that

$$N \subset I \subset Q \subset R \subset C$$

The definitions of N and I should be familiar to the reader. The definition of Q follows directly from the definition of a rational number.

Definition. *A* **rational number** *is one which can be put in the form* a/b *where* a *and* b *are integers and* b $\neq$ 0.

The definition of R may require a little more explanation. Since we have defined a rational number we are in a position to define an irrational number by a negation.

Definition. *An* **irrational number** *is one which cannot be expressed exactly as the quotient of two integers.*

The set of real numbers can then be defined as the set composed of all the rational and irrational numbers.* The definition given above in IV. is somewhat better, however, because the rational and irrational numbers can be expressed by decimal expansions.

Examples:

a) $3 = 3.000\ldots$
b) $1/2 = 0.5000\ldots$
c) $4/3 = 1.3333\ldots$
d) $\sqrt{2} = 1.41421\ldots$
e) $\pi = 3.141592653589793\ldots$

Note that if we let the set of real numbers, R, be the universe for our discussion, then the set of rationals and the set of irrationals are complements of each other:

$$\text{Set of irrationals} = Q' = R - Q$$

The definition of complex numbers may be new to the reader. We shall discuss such numbers in some detail in Section 38.

28. Operations. We shall use two operations to develop the real-number system. These are the operations of *addition*, "+", and *multiplication*, "·". The operation of multiplication is also denoted by parentheses, $(a)(b)$, and is understood when we write ab. Although we need only two operations to develop the real numbers we are not

---

*A more mathematically elegant definition of the real numbers depends upon sequences and limits, which are beyond the scope of this book. If you are interested, see E. J. Cogan, *Foundations of Analysis,* Englewood Cliffs: Prentice-Hall, Inc., 1962, p. 136.

discarding the other two (subtraction and division) which are used in arithmetic. As we shall see, subtraction and division can become a part of the operations of addition and multiplication through the idea of an *inverse*.

29. The Axioms of Real Numbers. We shall now consider the axioms which characterize the set of real numbers. These axioms further define the kinds of elements and operations being considered in the discussion of real numbers. In the statements of the axioms we shall let R represent the set of real numbers and letters of the alphabet represent the elements.

Axiom R_1. The Closure Axioms. *For any* $a \in R$ *and* $b \in R$, *we have* $(a + b) \in R$ *and* $a \cdot b \in R$.

These axioms state that the operations of addition and multiplication on real numbers each yield a unique real number. In other words, the sum of two real numbers is a real number and the product of two real numbers is a real number. We call this set of axioms the closure axioms and will say that the real numbers are closed under addition and multiplication.

Examples:

a) $2 + 3 = 5$ b) $2 \cdot 3 = 6$

c) $\dfrac{1}{2} + \dfrac{1}{3} = \dfrac{5}{6}$ d) $\dfrac{1}{2} \cdot \dfrac{1}{3} = \dfrac{1}{6}$

e) $1 + \sqrt{2} = 2.414\ldots$ f) $(\sqrt{2})(\sqrt{2}) = 2$

Axiom R_2. The Commutative Axioms. *For any* $a \in R$ *and* $b \in R$, *we have*

$$a + b = b + a \quad \text{and} \quad a \cdot b = b \cdot a$$

These axioms tell us that the order in which we add or multiply real numbers does not affect the results.

Examples:

a) $2 + 3 = 3 + 2 = 5$ b) $\dfrac{1}{2} \cdot 3 = 3 \cdot \dfrac{1}{2} = \dfrac{3}{2}$

Axiom R_3. The Associative Axioms. *For any* a, b, *and* c, *all elements of* R, *we have*

$$(a + b) + c = a + (b + c)$$
$$(a \cdot b) \cdot c = a \cdot (b \cdot c)$$

These axioms tell us that we may group the real numbers for addition and multiplication in any convenient manner. It is now possible to define the sum and product of three elements to be:

$$(a + b) + c = a + (b + c) = a + b + c$$
$$(a \cdot b) \cdot c = a \cdot (b \cdot c) = abc$$

This idea can be extended to the sum and product of any number of elements of R.

Examples:

a) $(7 + 3) + 2 = 10 + 2 = 12$ and $7 + (3 + 2) = 7 + 5 = 12$

b) $(3 \cdot 4) \cdot 5 = 12 \cdot 5 = 60$ and $3 \cdot (4 \cdot 5) = 3 \cdot 20 = 60$

The last two axioms permit us to rearrange the order of finding a sum or a product.

Illustration. Show that $a + b + c = c + b + a$.

Solution.
$$
\begin{aligned}
a + b + c &= (a + b) + c &&\text{[Definition]} \\
&= c + (a + b) &&\text{[Axiom } R_2] \\
&= c + (b + a) &&\text{[Axiom } R_2] \\
&= c + b + a &&\text{[Definition]}
\end{aligned}
$$

Axiom R_4. The Distributive Axioms. *For any* a, b, *and* c *in* R, *we have*

$$a \cdot (b + c) = a \cdot b + a \cdot c$$
$$(b + c) \cdot a = b \cdot a + c \cdot a$$

These axioms combine addition and multiplication, of which the reader has already made extensive use.

Examples:

a) $5 \cdot (2 + 6) = 5 \cdot 2 + 5 \cdot 6 = 10 + 30 = 40$

b) $3(x + y) = 3x + 3y$

c) $12x + 20y = 4(3x + 5y)$

d) $(6)(47) = 6(40 + 7) = 240 + 42 = 282$

The multiplication of polynomials makes extensive use of these axioms.

Illustration. Show that $(a + b)(c + d) = ac + ad + bc + bd$.

Solution.
$$
\begin{aligned}
(a + b)(c + d) &= (a + b)c + (a + b)d &&\text{[Axiom } R_4] \\
&= ac + bc + ad + bd &&\text{[Axiom } R_4] \\
&= ac + ad + bc + bd &&\text{[Definition]}
\end{aligned}
$$

If we consider the two statements of Axiom R_4 and the properties of Axiom R_3 we see that

$$a \cdot (b + c) = (b + c) \cdot a$$

so that the two statements express the same concept.

Axiom R_5. The Identity Axioms. *In the real-number system,* R, *there are two distinct elements,* $0 \in R$ *and* $1 \in R$, *such that for any* a $\in$ R *we have*

$$a + 0 = a \quad \text{and} \quad a \cdot 1 = a$$

The number *zero*, 0, is called the **identity element** for addition and the number *one*, 1, is called the **identity element** for multiplication. By using the commutative axioms we see that $a + 0 = 0 + a = a$ and $a \cdot 1 = 1 \cdot a = a$.

Axiom R_6. The Inverse Axioms. *For any* a $\in R$, *there exists an element of* R, *denoted by* $-$ a *such that*

$$a + (-a) = 0$$

Furthermore, if a $\in R$ (a $\neq 0$), *then there exists an element of* R, *denoted by* $1/a$, *such that*

$$a \cdot \frac{1}{a} = 1$$

The elements $-a$ and $1/a$ are called the **additive inverse** and **multiplicative inverse** of a, respectively. We also refer to $-a$ as the *negative* of a and to $1/a$ as the *reciprocal* of a. Axiom R_6 states that the result of combining by addition or multiplication any element with its inverse is the identity element for the particular operation used.

This completes the list of axioms for the real-number system.

30. Fundamental Theorems. In the next few sections we shall prove some of the fundamental properties of the real-number system. In this presentation we desire to show that the real-number system can be developed into a logical structure and to display the methods used to prove the theorems. We shall present only a few of the properties of the real numbers, a complete survey of the theory of numbers being a study in itself.

Since we are considering two operations, addition and multiplication, many of the theorems will make similar statements for each operation. Where appropriate, we shall mark theorems A for addition and M for multiplication. All elements belong to R.

Theorem 1A. Additive Cancellation Property. *If* a $+$ c $=$ b $+$ c, *then* a $=$ b.

Proof. Since $c \in R$, then $-c \in R$ by Axiom R_6.

$$
\begin{array}{ll}
(a + c) + (-c) = (b + c) + (-c) & \text{[Equality Axiom]} \\
a + [c + (-c)] = b + [c + (-c)] & \text{[Axiom } R_3\text{]} \\
a + 0 = b + 0 & \text{[Axiom } R_6\text{]} \\
a = b & \text{[Axiom } R_5\text{]}
\end{array}
$$

Theorem 1M. Multiplicative Cancellation Property. *If* ac $=$ bc *and* c $\neq 0$, *then* a $=$ b.

Proof. By Axiom R_6 there exists an element $1/c$, and we have

$$ac = bc \qquad \text{[Hypothesis]}$$

$$(ac)\frac{1}{c} = (bc)\frac{1}{c} \qquad \text{[Equality Axiom]}$$

$$a\left(c \cdot \frac{1}{c}\right) = b\left(c \cdot \frac{1}{c}\right) \qquad \text{[Axiom } R_3\text{]}$$

$$a \cdot 1 = b \cdot 1 \qquad \text{[Axiom } R_6\text{]}$$

$$a = b \qquad \text{[Axiom } R_5\text{]}$$

Definition.   *If, for a given property, there exists one and only one element with this property, the element is said to be* **unique.**

Theorem 2.   *The identity elements are unique.*
Proof.   Let b be any number in R such that

$$a + b = a$$

add $-a$:
$$-a + (a + b) = -a + a \qquad \text{[Axiom } E_4\text{]}$$
$$(-a + a) + b = -a + a \qquad \text{[Axiom } R_3\text{]}$$
$$0 + b = 0 \qquad \text{[Axiom } R_6\text{]}$$
$$b = 0 \qquad \text{[Axiom } R_5\text{]}$$

Since b was any number and $b = 0$, then 0 is unique.

Let b be any number in R such that

$$a \cdot b = a$$

multiply by $\frac{1}{a}$:
$$\frac{1}{a}(a \cdot b) = \frac{1}{a} \cdot a \qquad \text{[Axiom } E_5\text{]}$$

$$\left(a \cdot \frac{1}{a}\right) \cdot b = a \cdot \frac{1}{a} \qquad \text{[Axioms } R_2 \text{ and } R_3\text{]}$$

$$1 \cdot b = 1 \qquad \text{[Axiom } R_6\text{]}$$
$$b = 1 \qquad \text{[Axiom } R_5\text{]}$$

Since b was any number the identity element for multiplication is unique.

Theorem 3.   *The inverse elements are unique.* This theorem can also be stated in terms of equations.

  3A.   *If* $a + b = 0$, *then* $b = -a$.
  3M.   *If* $ab = 1$ *and* $a \neq 0$, *then* $b = 1/a$.

The proof of Theorem 3 is very similar to the proof of Theorem 2 and we leave it as an exercise for the reader.

Theorem 4A.   *If* $a = b$, *then* $-a = -b$.
Proof.
$$a = b \qquad \text{[Hypothesis]}$$
$$a + (-a) + (-b) = b + (-a) + (-b) \qquad \text{[Axiom } E_4\text{]}$$
$$0 + (-b) = 0 + (-a) \qquad \text{[Axioms } R_6, R_2, \text{ and } R_3\text{]}$$
$$-b = -a \qquad \text{[Axiom } R_5\text{]}$$
$$-a = -b \qquad \text{[Equality Axiom]}$$

Theorem 4M. *If* a = b, *then* 1/a = 1/b.
Proof.

$$a \cdot \frac{1}{a} = b \cdot \frac{1}{b} \qquad \text{[Both equal to 1]}$$

Since $a = b$:

$$b \cdot \frac{1}{a} = b \cdot \frac{1}{b} \qquad \text{[Substitution of equals]}$$

$$\frac{1}{a} = \frac{1}{b} \qquad \text{[Theorem 1M]}$$

Theorem 5.

5A. *If* a + x = b, *then* x = b + (− a).

5M. *If* ax = b *and* a ≠ 0, *then* x = b/a.

Proof. Given $a + x = b$, add $(-a)$ to both sides by the Equality Axiom E_4 and apply Axiom R_6 to obtain the result. Given $ax = b$, multiply both sides by the multiplicative inverse of a and apply Axiom R_6 to obtain the result.

Theorem 6. *The converse of Theorem 5 is true.* A direct statement of this theorem and its proof are left as an exercise.

Theorem 7. *The inverse of the inverse of an element is the element itself.*

Proof.

7A. The additive inverse of a is $-a$ and the additive inverse of $-a$ is $-(-a)$. Axiom R_6 states that $(-a) + [-(-a)] = 0$. If we add a to both sides we obtain

$$a + (-a) + [-(-a)] = 0 + a$$
$$0 + [-(-a)] = 0 + a \qquad \text{[Axiom } R_6\text{]}$$
$$-(-a) = a \qquad \text{[Axiom } R_5\text{]}$$

7M. The multiplicative inverse of a is $1/a$. Let us write the reciprocal of $1/a$ as $[1/a]^{-1}$. Axiom R_6 states that

$$\frac{1}{a} \cdot \left[\frac{1}{a}\right]^{-1} = 1$$

Multiply both sides by a to obtain

$$a \cdot \frac{1}{a} \cdot \left[\frac{1}{a}\right]^{-1} = a \cdot 1$$

Then, applying Axioms R_5 and R_6, we have

$$\left[\frac{1}{a}\right]^{-1} = a$$

Theorem 8. *The negative of zero is zero.*

Proof. Let $a = 0$ and $a = -0$ in Axioms R_5 and R_6 to obtain $0 + 0 = 0$ and $(-0) + 0 = 0$. Then by Axiom E_3 we have $-0 + 0 = 0 + 0$, which results in $-0 = 0$ by Theorem 1A.

31. The Number Zero. The identity element for addition, 0, has some special properties which are very important. Before stating the first property we shall clarify the mathematical use of the word "or." In mathematics the word "or" is always taken in the inclusive sense "and/or" so that the statement "*A* or *B*" means "either *A* or *B* or both *A* and *B*."

Property 1. *If* a $= 0$ *or* b $= 0$, *then* ab $= 0$.

Proof. Let
$$b = 0 = 0 + 0 \qquad \text{[Axiom } R_5]$$
$$a \cdot (0 + 0) = a \cdot 0 \qquad \text{[Axiom } E_5]$$
$$a \cdot 0 + a \cdot 0 = a \cdot 0 \qquad \text{[Axiom } R_4]$$
$$= a \cdot 0 + 0 \qquad \text{[Axiom } R_5]$$
$$a \cdot 0 = 0 \qquad \text{[Theorem 1A]}$$

The same argument holds if $a = 0$. If both a and b equal 0, repeat the above argument with a replaced by 0.

Property 2. *The converse of Property* 1. *If* ab $= 0$, *then* a $= 0$ *or* b $= 0$.

Proof. If $a = 0$, then $ab = 0$ by Property 1. If $a \neq 0$, then the multiplicative inverse of a exists and we have $1/a \cdot ab = 1/a \cdot 0$, so that $1 \cdot b = 0$, or $b = 0$, by Axioms R_3 and R_6.

Property 3. *The product* $0 \cdot 1/c$ *equals* 0. This is a special case of Property 1 with $a = 0$ and $b = 1/c$.

Property 4. *The multiplicative inverse of* 0 *does not exist.*

Proof. Suppose the multiplicative inverse of 0 did exist; call it c. Then by Axiom R_6 we would have $0 \cdot c = 1$, but this contradicts Property 1, which proved that $0 \cdot c = 0$.

Since Axiom R_6 defined

$$a \cdot \frac{1}{a} = \frac{a}{a} = 1$$

it is natural to ask, "What is 0/0?" even though the reciprocal of 0 does not exist. The answer is dependent upon how this expression happened to occur in the discussion. For the present we shall simply say that this expression is *indeterminate*. We shall discuss "division" by 0 in the next section and also in Section 71.

32. Subtraction and Division. We have developed the real-number system under the two operations of addition and multiplication. During our study of arithmetic we had two additional operations, subtraction and division. These two operations may be defined as the inverses of addition and multiplication.

Definition. *The difference of* a *and* b, *or* b *subtracted from* a, b − a, *is equivalent to* a + (− b).

Definition. *The quotient of* a *by* b *where* b ≠ 0, *written* $\dfrac{a}{b}$, *is equivalent to* a · $\left(\dfrac{1}{b}\right)$.

All of the properties with which we are already familiar hold for these definitions of subtraction and division, and the axioms are valid. It is cumbersome, however, to perform these operations with the notation of inverses. Consequently, we have adopted the simpler notation $a - b$ and $a \div b$.

This notation causes some complication. First, the minus sign "−" now has two meanings: "the negative of a" and "the difference of b and a." The two meanings do not cause too much trouble if we adopt additional rules and clearly define the subtraction and product of negative numbers. Furthermore, subtraction and division no longer obey the commutative and associative axioms, for $a - b \neq b - a$ and $a \div (b \div c) \neq (a \div b) \div c$. Consequently, there are no identity elements for subtraction and division, and no inverses. Since $(a - b) - c \neq a - (b - c)$, we need to keep the order of subtraction in a specified sequence and also to define the operation of removing parentheses when preceded by a minus sign. These rules are well known to the reader but it may be advantageous to review the discussion in Chapter I.

In the last section we proved that the multiplicative inverse of the number 0 does not exist. Consequently, division by 0 is not defined by our definition. Since we are unable to formulate a definition for division by 0 which is consistent with our axioms for real numbers, this operation is not permissible. We shall therefore say that *any algebraic expression in the real-number system is undefined for those values of the letters which make a denominator 0.*

Some mathematicians prefer to list the definitions for subtraction and division as axioms for the real numbers. Either method yields the four basic operations.

33. Order. In Section 13 we listed three axioms of order which permit comparison of quantities. We shall now apply these axioms to the real-number system.

Definition. *A real number* a *is* **positive** *if* a > 0 *and* **negative** *if* a < 0.

On the basic of this definition the number 0 is neither positive nor negative; it is frequently said that 0 separates the positive and negative numbers. We shall now add a fourth axiom to the three given in Section 13.

Axiom 0_4. The Multiplication Axiom. *If* a, b, *and* c *are elements of* R *such that* a > b *and* c > 0, *then* ac > bc.

Examples:

a) Let $c = 2$ and since $5 > 3$, then $(5)(2) > (3)(2)$ or $10 > 6$.

b) Since $5 > 3$ we have $5/2 > 3/2$ by letting $c = 1/2$.

Theorem 1. *If* a *and* b *are elements of* R, *then*

   A. a > b *if and only if* $-$ a < $-$ b.

   B. a < b *if and only if* $-$ a > $-$ b.

Proof.	**A.**	**B.**	Reason
Part I.	$a > b$ $a + c > b + c$	$a < b$ $a + c < b + c$	Hypothesis Axiom 0_3

Let $c = (-a) + (-b)$ and apply Axioms R_2, R_3, and R_6.

$0 + (-b) >$
      $0 + (-a)$
 $-b > -a$
 $-a < -b$
	$0 + (-b) <$
	      $0 + (-a)$
	 $-b < -a$
	 $-a > -b$
		Axioms R_2, R_3, R_6
		
		 Axiom R_5
		 Definitions of
		          $<$ and $>$

$-a < -b$	$-a > -b$	Hypothesis

Part II. Repeat the steps of Part I in reverse order.

If we let $b = 0$ in Theorem 1 above and recall that $0 = -0$ (Theorem 8, Section 30) we have proved an important corollary.

Corollary.   **A.** *If* a > 0, *then* $-$ a < 0.

            B. *If* a < 0, *then* $-$ a > 0.

Examples:

a) If $a = 2$, we have $2 > 0$ and $-2 < 0$.

b) If $a = -2$, we have $-2 < 0$ and $-(-2) = 2 > 0$.

This corollary emphasizes the difference between "negative number" and "negative of a number." We have formulated the definition so that a *negative number* is one that is less than 0 but the *negative of a number* can be a positive number. In fact, the negative of a number is a positive number if the number itself is a negative number; Example (b) illustrates the point. Notice that this is consistent with our idea that the product of two negative numbers is a positive number. We shall prove this last statement but first we shall need another theorem.

Theorem 2. *If* $a > b$ *and* $c < 0$, *then* $ac < bc$.

Proof. Since $c < 0$, we have $-c > 0$ by Theorem 1. Now Axiom O_4 permits us to multiply both sides of the inequality $a > b$ by $(-c)$ to obtain $a(-c) > b(-c)$ or $-ac > -bc$. Then by Theorem 1A, we have $ac < bc$.

Example: Since Since $4 > 3$ and $-2 < 0$, we have $-8 < -6$.

> **Illustration.** Show that if $a < 0$ and $b < 0$, then $ab > 0$. (The product of two negative numbers is a positive number.)
>
> *Solution.* Start with the given property $a < 0$ and multiply both sides by b; then, since $b < 0$, we have, by Theorem 2, $ab > 0 \cdot b$, or $ab > 0$ by Property 1, Section 31.

Theorem 3. *If* $a > b$ *and* $c > d$, *then* $a + c > b + d$.

Proof.

$a > b$	and	$c > d$	[Given]
$a + c > b + c$	and	$b + c > b + d$	[Axiom O_3]
$\therefore$ $a + c > b + d$			[Axiom O_2]

Examples:

a) Since $5 > 3$ and $4 > 2$, we have $9 > 5$.

b) Since $2 < 3$ and $4 < 5$, we have $6 < 8$.

Theorem 4. *If* a, b, c, *and* d *are positive numbers such that* $a > b$ *and* $c > d$, *then* $ac > bd$.

Proof.

$a > b$	and	$c > d$	[Given]
$ac > bc$	and	$bc > bd$	[$c > 0, b > 0$, Axiom O_4]
$\therefore$ $ac > bd$			[Axiom O_2]

Examples:

a) Since $5 > 3$ and $4 > 2$, we have $20 > 6$.

b) Since $2 < 3$ and $4 < 5$, we have $8 < 15$.

These axioms and theorems form the basis for operations on **inequalities** involving elements from the real-number system. We shall restate these operations in terms that may be more familiar to the reader. Two inequalities are said to have the *same sense* if their symbols for inequality point in the same direction and the *opposite sense* if the symbols point in opposite directions.

Properties of Operations:

I. The sense of an inequality, $a > b$, is *not* changed

    a) if both members are *increased* by the *same* number,

b) if both members are *diminished* by the *same* number,

c) if both members are *multiplied* by the same *positive* number,

d) if both members are *divided* by the same *positive* number,

e) if both members are *positive* and are raised to the *same* power,

f) if the corresponding members of an inequality in the *same sense* are added,

g) if the corresponding members of an inequality in the *same sense* are multiplied.

II. The sense of an inequality is *reversed*

a) if both members are multiplied by the *same negative* number,

b) if the *reciprocals* (multiplicative inverses) of two nonzero elements of like sign are compared.

The following series of illustrations will prove several properties of the real numbers; others will be proved in the exercises.

Illustrations.

1) Prove that the identity element for multiplication is positive; i.e., that $1 > 0$.

Solution. By Axiom 0_1, we have $1 < 0$, $1 = 0$, or $1 > 0$. By Axiom R_5, we have $1 \neq 0$. Suppose $1 < 0$, then $-1 > 0$ by Corollary B. Since $-1 > 0$, $(-1)(-1) > 0(-1)$ by Axiom 0_4. Then $1 > 0$ by the Illustration on page 52. But this contradicts the assumption that $1 < 0$; consequently, we have $1 > 0$.

2) (Property II [b]) Show that if a and b are positive and $a > b$, then $(1/a) < (1/b)$.

Solution. Assume that $(1/a) > (1/b)$.

$$\text{Since } a > 0, a\left(\frac{1}{a}\right) > a\left(\frac{1}{b}\right) \qquad [\text{Axiom } 0_4]$$

$$\text{Since } b > 0, 1(b) > a\left(\frac{1}{b}\right)b \qquad [\text{Axioms } 0_4 \text{ and } R_6]$$

Applying Axiom R_6 again we have

$$b > a$$

but this contradicts the hypothesis that $a > b$; consequently, our assumption is wrong and $(1/a) < (1/b)$.

3) Show that if $a \in R$ and $a \neq 0$, then

a) $a^2 > 0$ b) $\left(\frac{1}{a}\right) > 0$ if $a > 0$ c) $\left(\frac{1}{a}\right) < 0$ if $a < 0$

Solution. (a) If $a > 0$, then $a^2 > 0$ by Axiom 0_4. If $a < 0$, then $a^2 > 0$ by the Illustration on page 52. (b) Since (a) is true for all elements of R

except 0, it is true for $\dfrac{1}{a}$, and $\left(\dfrac{1}{a}\right)^2 > 0$. Then by Axiom 0_4, since $a > 0$,

we have $a\left(\dfrac{1}{a}\right)^2 = \dfrac{1}{a} > 0$. (c) Since $\left(\dfrac{1}{a}\right)^2 > 0$ and $a < 0$, we have

$a \cdot \left(\dfrac{1}{a}\right)^2 = \dfrac{1}{a} < 0$ by Theorem 2.

There are many more interesting properties of the real numbers; we cannot discuss them all. However, we have been referring to the positive integers and the negative integers and it may interest the reader to know that we can now show that the integers $1, 2, 3, 4, 5, \ldots$ are positive and that the integers $-1, -2, -3, -4, \ldots$ are negative.

Illustration. Show that $n > 0$ if $n = 1, 2, 3, \ldots$ and that $n < 0$ if $n = -1, -2, -3, \ldots$.
Solution. In Illustration 1 on page 53 we proved that $1 > 0$. The number $2 = 1 + 1$ by definition. If we add 1 to both sides of $1 > 0$, we obtain

$$1 + 1 = 2 > 0 + 1 = 1 > 0$$

Therefore, $2 > 0$ by Axiom 0_2.
Since $3 = 2 + 1$ we can proceed in the same manner to obtain

$$3 = 2 + 1 > 0 + 1 = 1 > 0$$

The process can be continued for all n.
Since $n > 0$, we have $-n < 0$ by Theorem 1.

Definition. *The element* c *is said to be* **between** a *and* b *if*
$a > c > b$.

Definition. *Set* A $= \{a_i | i = 1, 2, 3, \ldots \}$ *is said to be* **dense** *if for every pair of elements* a_i, a_j *there exists at least one element* a_k *which is between* a_i *and* a_j.

Theorem 5. *The set of real numbers* R *is dense.*
Proof. Let $a \in R$ and $b \in R$ such that $a \neq b$; then either $a > b$ or $a < b$. Let us take $a > b$. Then

$$a + a = 2a > a + b \qquad\qquad \text{[Axiom } 0_3]$$

Since $\dfrac{1}{2} \in R$ and $\dfrac{1}{2} > 0$ we have

$$a > \dfrac{a + b}{2} \qquad\qquad \text{[Axiom } 0_4]$$

Since $b \in R$ we can add b to $a > b$ to obtain

$$a + b > b + b = 2b$$

$$\frac{a + b}{2} > b \qquad \text{[Axiom } O_3\text{]}$$

$$\therefore \qquad a > \frac{a + b}{2} > b \qquad \text{[Axiom } O_4\text{]}$$

Now $\frac{a + b}{2} \in R$ by the Closure Axiom R_1 and we have proved the theorem.

Theorem 5 is a very important theorem concerning the real numbers. Since the element between a and b is an element of R, the theorem says that there is an element $d \in R$ such that $a > d > (a + b)/2 > b$. The theorem can be applied again and again so that in a more general form we have the following statement:

There are infinitely many real numbers between any pair of distinct real numbers.

Notice that this theorem is not true if we limit ourselves to the set of integers, since there is no integer between two consecutive integers.

There are two more properties of order which we shall include among the axioms. We shall first need a definition.

Definition. *If the elements of a set are ordered so that there is a first element a_1, a second element a_2, a third element a_3, etc., then the ordered set is called a* **sequence.**

Examples:
a) The natural numbers, 1, 2, 3, . . . , form a sequence.
b) The set $\{b_i, i = 1, 2, 3, \ldots\}$ is a sequence.

Axiom O_5. The Axiom of Archimedes. *If a and b are positive real numbers and* $a < b$, *then there is a positive integer* n *such that* $na > b$.

Example: $2 < 5$ and with $n = 3$ we have $3(2) = 6 > 5$. Note that we can select n to be the integer that is larger than the quotient of b and a.

Axiom O_6. Nesting. *If* $\{a_i\}$ *and* $\{b_i\}$, $i = 1, 2, 3, \ldots$, *are two sequences of real numbers such that*

(a) $a_i \leq a_{i+1}$ (b) $b_i \geq b_{i+1}$ (c) $a_n \leq b_n$ *and* (d) $b_n - a_n < 10^{-n}$

for every $i \in N$ *and* $n \in N$ *then there is one and only one real number* c *such that* $a_n < c < b_n$ *for every natural number* n.

34. Absolute Value. The absolute value of a real number is an important concept of the real-number system. Although we mentioned it in Chapter I, let us now give a rigorous definition of this idea.

Definition. *The absolute value of a real number* x *is defined as*
$$|x| = x \text{ if } x > 0 \qquad |x| = -x \text{ if } x < 0 \qquad |x| = 0 \text{ if } x = 0$$

Examples:

$$|3| = 3 \qquad |-3| = -(-3) = 3 \qquad |0| = 0$$

We emphasize that the absolute value of a real number is always positive.

Theorem 1. *For any* $a \in R$, $-|a| \le a \le |a|$.

Proof. If $a = 0$, then $|a| = 0$, and since $-0 = 0$ we have $0 = 0 = 0$ and the theorem is satisfied. If $a > 0$, then $-a < 0$ by Theorem 1, Section 33. Furthermore, $|a| = a$ and $-a = -|a| < 0 < a = |a|$. Thus the theorem is satisfied. If $a < 0$, then $0 < -a$ by Theorem 1, Section 33. Furthermore, $|a| = -a$, so that $-|a| = a$ and $0 < |a|$. Thus we have $-|a| = a < 0 < |a|$ and the theorem is satisfied.

The reader is already familiar with the process of taking a root and in Chapter I we defined the symbol $\sqrt{}$ to mean the nonnegative square root. We can write this in terms of the absolute value.

Definition. $\sqrt{a^2} = |a|$

Examples:

a) $\sqrt{4} = \sqrt{2^2} = 2$
b) $\sqrt{4} = \sqrt{(-2)^2} = 2$
c) $-\sqrt{4} = -\sqrt{2^2} = -|2| = -2$

Theorem 2. *The absolute value of the product (or quotient) of two real numbers is equal to the product (or quotient) of their absolute values:*

$$|ab| = |a| \cdot |b| \qquad \text{and} \qquad \left|\frac{a}{b}\right| = \frac{|a|}{|b|}, \quad b \ne 0$$

Proof. We shall use the above definition and the properties of radicals to write

$$|ab| = \sqrt{(ab)^2} = \sqrt{a^2b^2} = \sqrt{a^2}\sqrt{b^2} = |a| \cdot |b|$$
$$\left|\frac{a}{b}\right| = \sqrt{\frac{a^2}{b^2}} = \frac{\sqrt{a^2}}{\sqrt{b^2}} = \frac{|a|}{|b|}$$

Examples:

a) $|6| = |2||3|$

b) $\left|\frac{3}{5}\right| = \dfrac{|3|}{|5|}$

Theorem 3. Triangle Inequality. *The absolute value of the sum of two real numbers is less than or equal to the sum of their absolute values:*

$$|a + b| \leq |a| + |b|$$

Proof. By Theorem 1 we have

$$-|a| \leq a \leq |a| \qquad \text{and} \qquad -|b| \leq b \leq |b|$$

Add these inequalities to obtain

$$-|a| - |b| \leq a + b \leq |a| + |b|$$

If $a + b > 0$, then by definition we have $a + b = |a + b|$ and we can substitute to obtain the statement of the theorem. If $a + b < 0$, then $|a + b| = -(a + b)$ by definition. Since $-(|a| + |b|) \leq a + b$, we have $-(a + b) \leq |a| + |b|$ by Theorem 2, Section 33 and, consequently,

$$-(a + b) = |a + b| \leq |a| + |b|$$

35. Exercise III.

1. Find the additive inverses of 1, 3, -6, 1/3, and $-\sqrt{2}$

2. Find the multiplicative inverses of 1, 7, 1/3, -3, 0, and $\sqrt{2}$.

3. Draw a Venn diagram with the real numbers as the universe. Show the subset relationship among the sets of positive integers, negative integers, positive fractions, negative fractions, zero, the positive rational numbers, the negative rational numbers, and the rational numbers.

4. True or False:

   a) The set of integers is closed under subtraction.

   b) The set of integers is closed under division.

   c) $a \div (b + c) = (a \div b) + (a \div c)$

   d) If $a < b$ and $c < d$, then $a - c < b - d$ for all $a, b, c,$ and d.

   e) The set of real numbers is closed under division.

   f) If $a > 1$, then $a < a^2$.

   g) For all $a \in R$, we have $|a| = a$.

   h) If $a > b$ and $b > c$, then $c - x < a - x$.

   i) If $a > b$ and $c < 0$. then $bc > ac$.

   j) If $a > b$ and $c > d$, then $b + d < a + c$.

5. Prove that if $a \in R$, then $a^2 > 0$.

6. Show that $b + (a - b) = a$.

7. If $a \neq 0$, show that $a(b/a) = b$.

8. If $b \neq 0$, show that $a/b = c$ if and only if $a = bc$.

9. If $ax = b$, $a \neq 0$, show that $x = b/a$.

10. Prove that if $a > 0$, $b > 0$, and $a > b$, then $a^2 > b^2$.

36. Some Special Properties. In this section we shall consider a few interesting properties of the real numbers. We have already defined the set of real numbers as those that can be expressed by a decimal expansion and the subset Q of rational numbers as those that can be expressed as the quotient of two integers. We can show, by performing the division, that

$$\frac{1}{3} = 0.3333\ldots$$

$$\frac{1}{7} = .142857142857\ldots$$

and notice that certain groups of digits repeat themselves, (3) and (142857). The repeating digits do not necessarily start with the first digit; e.g.,

$$\frac{11}{6} = 1.8333\ldots$$

For ease of writing we shall place a bar over the group of digits which is repeated indefinitely:

$$1.\overline{3} = 1.3333\ldots$$
$$6.4\overline{35} = 6.4353535\ldots$$

Theorem 1. *Every repeating decimal expansion is a rational number.*
Proof. Let us write the repeating decimal in the form

$$N = a_0 . a_1 a_2 \ldots a_k \overline{b_j \ldots b_r}$$

where the a_i $(i = 0, 1, \ldots, k)$ and b_j $(j = 1, \ldots r)$ represent digits and the b_j are repeated. Multiply both sides by 10^k to place the decimal point after a_k:

$$10^k N = a_0 a_1 \ldots a_k . \overline{b_1 \ldots b_r}$$

and rewrite this number as

$$M = M_1 . \overline{b_1 \ldots b_r}$$

Now multiply this number by 10^r to place the first group of repeating digits to the left of the decimal point, and then subtract M:

$$10^r M = M_1 b_1 \ldots b_r . \overline{b_1 \ldots b_r}$$
$$M = M_1 . \overline{b_1 \ldots b_r}$$

$$10^r M - M = M_1 b_1 \ldots b_r - M_1$$
$$(10^r - 1)M = M_1 b_1 \ldots b_r - M_1$$

The right-hand side will be an integer; call it M_2 and solve for M:

$$M = \frac{M_2}{10^r - 1}$$

so that M is a rational number; and since $N = M/10^k$ we have shown that the repeating decimal is a rational number.

The proof of the theorem establishes the procedure for finding the rational number for a given repeating decimal expansion.

Illustrations.
1) Find the rational number represented by $1.\overline{23}$.
Solution.　Multiply by $10^2 = 100$ to obtain

$$100N = 123.\overline{23}$$

Subtract the original number:

$$100N - N = 123.\overline{23} - 1.\overline{23} = 122$$
$$99N = 122$$
$$N = \frac{122}{99}$$

2) Find the rational number represented by $2.3\overline{45}$.
Solution.　First multiply by 10 to move the repeating group immediately to the right of the decimal point:

$$10N = 23.\overline{45}$$

Now multiply by $10^2 = 100$ to obtain

$$1000N = 2345.\overline{45}$$

and subtract $10N = 23.\overline{45}$ to obtain

$$990N = 2322$$
$$N = \frac{1161}{495}$$

The set of integers I can be separated into two parts, the even integers and the odd integers, and be written

even integers: $a = 2n$ and $a^2 = 4n^2$

odd integers: $a = 2n + 1$ and $a^2 = 4n^2 + 4n + 1$

where $n \in I$; that is, $n = 0, \pm 1, \pm 2, \ldots$.

Theorem 2. *If* a^2 *is divisible* by* 2, *then* a *is divisible by* 2.

Proof. Since a^2 is divisible by 2 it must have the form $4n^2$ because $4n^2 + 4n + 1$ is not divisible by 2. Hence $a = 2n$, and a is divisible by 2.

Theorem 3. $\sqrt{2}$ *is not a rational number.*

Proof. The theorem will be proved by contradiction. Suppose that p/q is a rational number such that p and q have no common factors and $p^2/q^2 = 2$, so that $p^2 = 2q^2$. Then p^2 is divisible by 2 and thus, by Theorem 2, p is divisible by 2 and we can write $p = 2n$. Then $p^2 = 4n^2 = 2q^2$ or $2n^2 = q^2$. Hence q^2 is divisible by 2 and, by Theorem 2, q is divisible by 2. However, if p and q are both divisible by 2, they have 2 as a common factor, contradicting our assumption.

In a similar manner we can prove that $\sqrt{3}$, $\sqrt{5}$, etc. are irrational numbers. It is much more difficult, however, to prove that π is irrational. The reader can construct many other irrational numbers. Any decimal expansion is an irrational number providing no group of digits is repeating. Thus, if we form a decimal expansion by a rule which insures that no group of digits repeats, then we have an irrational number. Consider, for example, the number formed by writing after the decimal point 2, then 3, then 2, then two 3's, then 2, then three 3's, etc.:

$$.232332333233332333332\ldots$$

We know from the definition that 3/6 is a rational number and that 1/2 is a rational number. We also know from arithmetic that these two numbers are equal. Let us formulate and prove the theorem for equality of two rational numbers.

Theorem 4. *The rational numbers* a/b *and* c/d *are equal if and only if* ad = bc.

Proof. Part I. If $a/b = c/d$, then by Axiom E_5 we have

$$bd\left(\frac{a}{b}\right) = bd\left(\frac{c}{d}\right)$$

---

*"Divisible" means "exactly divisible"; i.e., the quotient $\in I$.

and by Axioms R_3, R_6, and R_5 we have

$$b \cdot \frac{1}{b} \cdot d \cdot a = b \cdot d \cdot \frac{1}{d} \cdot c$$

$$ad = bc$$

Part II. If $ad = bc$, then by Axiom E_5 we have

$$\frac{ad}{bd} = \frac{bc}{bd}$$

and by Axioms R_6 and R_5 we have

$$\frac{a}{b} = \frac{c}{d}$$

The set of natural numbers N has a most important characteristic. We shall state it in the form of an axiom.

Axiom N_1. Axiom of Mathematical Induction. *If* S *is a subset of the natural numbers,* $S \subset N$, *such that*
a) $1 \in S$
b) If $k \in S$, *then* $k + 1 \in S$
then S *is the set of all natural numbers,* $S = N$.

The idea expressed by this axiom is as follows: to form a subset S of the natural numbers one starts with the number 1 and then chooses a second element by the property that if S contains 1 then it contains $1 + 1$. This gives us the element 2. Then we repeat the process: if $2 \in S$, then $2 + 1 = 3 \in S$. We construct the set S by continuing this process. In so doing, however, we are generating exactly the set of natural numbers N; consequently $S = N$.

This axiom is used to prove many statements. A proof by mathematical induction consists of two steps:

1. Verify the statement for some $n \in N$.
2. Assume the statement for $n = k$; then prove it for $n = k + 1$. If both of these steps can be accomplished then the Axiom of Mathematical Induction is satisfied and the statement is true for *all n.*

Illustration. Prove that the sum of the first n natural numbers is $1/2\,(n)(n + 1)$:

$$1 + 2 + 3 + 4 + \ldots + n = \frac{1}{2}n(n + 1)$$

Solution. Step 1. Let $n = 1$. Then on the left-hand side we have the first term 1 and on the right-hand side we have $1/2\,(1)(1+1)=1$. Thus we have $1=1$ and the statement is true for $n=1$. We can now go to Step 2; however, let us check the statement for $n=2$ as a matter of interest.

$$1+2=\frac{1}{2}(2)(2+1) \text{ or } 3=3$$

Step 2.   Assume that the statement is true for $n=k$.

$$1+2+3+\ldots+k=\frac{1}{2}(k)(k+1)$$

Now we wish to prove that the statement is true for $n=k+1$. Let us therefore add the next term to the left-hand side and the same quantity to the right-hand side:

$$1+2+3+\ldots+k+(k+1)=\frac{1}{2}k(k+1)+(k+1)$$

Let us simplify the right-hand side:

$$\frac{1}{2}k(k+1)+(k+1)=(k+1)\left(\frac{1}{2}k+1\right)$$
$$=(k+1)\left(\frac{1}{2}\right)(k+2)$$
$$=\frac{1}{2}(k+1)(k+2)$$

If we substitute $n=k+1$ into the given statement, we obtain

$$1+2+3+\ldots+(k+1)=\frac{1}{2}(k+1)(k+2)$$

but this is what we obtained above. Consequently, we have shown that the statement is true for $n=k+1$.

Thus the statement satisfies the conditions of Axiom N_1 and is true for all n. Notice that in performing Step 2 we first do what the statement says to do for the next term and then we show that the result is the same as that obtained by a direct substitution of $n=k+1$ into the statement.

37. Geometric Representation.

Consider a straight line of unlimited length and choose a point O, which we shall call the **origin.** Place the line horizontally and let the direction be chosen *positive* if taken to the right of the origin and *negative* if taken to the left of it. Now choose an arbitrary length as the unit of measure and "lay off" successive units in both directions from the origin. Label the origin O, the markings to the right 1, 2, 3, 4, . . . , and the markings to the left -1, -2, -3, ... We have constructed a number scale as shown in Figure 5.

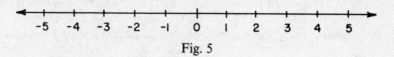

Fig. 5

On such a number scale, one can locate any real number whether it be rational or irrational. This statement is usually a part of a theorem. However, the proof of the theorem is beyond the scope of this book so we shall state it as a property and explain it.

Geometric Property. *There is a one-to-one correspondence between the points on a straight line of unlimited length and the elements of the set of real numbers.*

We have already shown a method for locating the set of integers. Since any line segment can be divided into n equal parts by a geometric construction, we can divide the segment from O to 1 into any given part and locate the fraction $1/n$. This construction can be made for any of the line segments between consecutive integers. It is thus that we locate all the rational numbers. The location of an irrational number is not so simple. They may be located only as accurately as the distance given by their decimal approximation can be measured. The converse—given a point on the line, find the associated number—is again a matter of approximation.

This method of representing the real-number system preserves the order of the real numbers. A larger number is always to the right of a smaller number and a smaller number is always to the left of a larger number. Test this statement by comparing the positions of the numbers in Figure 5.

The length of the line segment from point x_1 to point x_2 is called the **distance** between the two points. This distance is taken to be positive and is given by the absolute value of the difference of the two numbers represented by the points, $|x_1 - x_2|$. Since we use the absolute value, the order in which we choose the numbers to find the difference is immaterial.

Illustrations.
1) Find the distance between $P_1(5)$ and $P_2(3)$. [$P(x)$ denotes the point represented by the number x.]
Solution. $d = |5 - 3| = |3 - 5| = 2$
2) Find the distance between $P_1(5)$ and $P_2(-3)$.
Solution. $d = |5 - (-3)| = |(-3) - 5| = 8$
Check this by measuring the distance from (-3) to 5 on Figure 5.

The segment whose left end-point is a and whose right end-point is b is called the **interval** from a to b and is denoted by $[a, b]$. Since this line segment also represents the set of real numbers between a and b, we have

$$[a, b] = \{x | x \in R \text{ and } a \leq x \leq b\}$$

Note that in this definition the end-points a and b are included. When this is the case we call $[a, b]$ the **closed interval** from a to b. If one or both end-points are not included, the interval is said to be an **open interval.** Note the use of the brackets to indicate open and closed intervals.

$$]a, b] = \{x | x \in R \text{ and } a < x \leq b\}$$
$$[a, b[= \{x | x \in R \text{ and } a \leq x < b\}$$
$$]a, b[= \{x | x \in R \text{ and } a < x < b\}$$

This new notation eases our operations with the set of real numbers. *Examples:*
a) $[1, 2] \cup [2, 4] = [1, 4]$
b) $[3, 5] \cap [4, 6] = [4, 5]$
c) $[3, 6[\cup]1, 4] =]1, 6[$
d) $] - 1, 4] \cap]0, 7[=]0, 4]$
e) $] - 3, 7[\cap [- 2, 9] = [- 2, 7[$
f) $[1, 6]' = \text{complement of } [1, 6] =] - \infty, 1[\cup]6, + \infty[$

38. Complex Numbers. In Section 24 we defined the concept of an ordered pair. We shall use this concept to extend our number system. This extension is necessary because the real-number system is not closed under the operation of taking a root, since we cannot find a real number equal to the square root of a negative number.

Definition. *A* **complex number** *is an ordered pair of real numbers,* (a, b).

The arithmetic of these numbers can be formulated by the definition of operations.

Equality. *Two complex numbers,* (a, b) *and* (c, d), *are said to be* **equal** *if and only if* a = c *and* b = d.

Addition. *The* **sum** *of two complex numbers,* (a, b) *and* (c, d), *is a complex number whose first element is* a + c *and second element is* b + d:

$$(a, b) + (c, d) = (a + c, b + d)$$

Example:

$$(2, 5) + (3, -2) = (5, 3)$$

Multiplication. *The* **product** *of two complex numbers,* (a, b) *and* (c, d), *is a complex number whose first element is* ac − bd *and second element is* bc + ad:

$$(a, b) \cdot (c, d) = (ac - bd, bc + ad)$$

Example:

$$(2, 5) \cdot (3, 4) = (- 14, 23)$$

If $b = 0$ and $d = 0$ so that the complex numbers are $(a, 0)$ and $(c, 0)$ the arithmetic is exactly the same as that for the real numbers. We therefore say that $(a, 0)$ is a real number and call this real number the **real part** of the complex number (a, b).

If $a = 0$ and $c = 0$, we see that

$$(0, b) \cdot (0, d) = (- bd, 0)$$

which is a real number. This is an important characteristic of complex numbers. In particular, if we multiply such a number by itself, we have

$$(0, b) \cdot (0, b) = (- b^2, 0)$$

so that

$$\sqrt{(- b^2, 0)} = (0, b)$$

Consequently in the complex-number system we can find a number that is the square root of a negative real number. The number $(0, b)$ is called a **pure imaginary number** and is also called the **imaginary part** of the complex number (a, b).

Let us consider the special pure imaginary number $(0, 1)$. We see that

$$(0, 1) \cdot (0, 1) = (0, 1)^2 = (- 1, 0)$$

The number $(0, 1)$ is usually* denoted i, so that we have $i^2 = - 1$. This provides us with a second notation for a complex number:

$$(a, b) = a + bi$$

This is the more familiar notation used in Chapter I. Thus, instead of writing a complex number as an ordered pair, we can write it as a binomial following the rules for the arithmetic of polynomials. In this form it is important to remember that $i^2 = - 1$, $i^3 = - i$, and $i^4 = 1$.

A complex number is said to be the **conjugate** of another complex number if the two numbers differ only in the sign of the imaginary part.

---

*Electrical engineers call it j since i is used to denote current.

Examples:
a) The conjugate of $(2, 3)$ is $(2, -3)$.
b) The conjugate of $3 + 4i$ is $3 - 4i$.
c) The conjugate of $3 - 4i$ is $3 + 4i$.
d) The conjugate of $2i - 3$ is $-3 - 2i$.

Notice that if we write a complex number in the form $a + bi$, then $bi + a$ is the same number; the order in which we write the binomial does not specify the real and imaginary parts. On the other hand, if the complex number is written as an ordered pair (a, b), then the first element is always associated with the real part and the second element with the imaginary part.

The division of two complex numbers yields a complex number. To perform the arithmetic we employ the idea of the conjugate. Write the division in the form of a fraction and multiply the numerator and denominator by the conjugate of the denominator.*

Illustration. Find $(3 - 2i) \div (2 + i)$.
Solution:

$$\frac{3 - 2i}{2 + i} = \frac{3 - 2i}{2 + i} \cdot \frac{2 - i}{2 - i} = \frac{4 - 7i}{4 + 1} = \frac{4}{5} - \frac{7}{5}i$$

The number system we shall use in this book is represented in Figure 6.

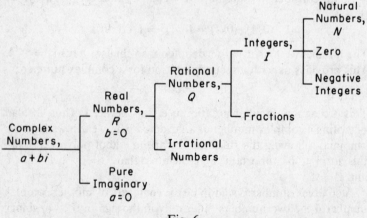

Fig. 6

---

*Compare this operation with that of "rationalizing the denominator."

Summary of Symbols	
$N = \{1, 2, 3, 4, \ldots\}$ $I = \{\ldots, -2, -1, 0, 1, 2, 3, \ldots\}$	0 = Identity element for addition 1 = Identity element for multiplication
$Q = \{x \mid x = \dfrac{a}{b}, a \in I, b \in I, b \neq 0\}$	$-a$ = Additive inverse of a
$R = \{x \mid x = d_0 d_1 d_2 d_3 \ldots\}$	$\dfrac{1}{a}$ = Multiplicative inverse of a
$C = \{x \mid x = a + bi, a \in R, \\ b \in R, i^2 = -1\}$	$\|a\|$, Absolute value of a
$R_1, R_2, R_3, R_4, R_5, R_6 =$ Axioms of Operations	$a + bi = (a, b)$, a complex number
$O_1, O_2, O_3, O_4, O_5, O_6 =$ Axioms of Order	i, Imaginary unit such that $i^2 = -1$

39. Exercise IV.

1. Find the rational number represented by the following:

    a) $1.\overline{3}$ b) $2.\overline{6}$

    c) $0.2\overline{45}$ d) $3.\overline{1}$

    e) $1.32\overline{45}$ f) $12.\overline{5}$

2. Use the axiom of mathematical induction to prove the following:

    a) $1 + 3 + 5 + 7 + \ldots + (2n - 1) = n^2$

    b) $2 + 4 + 6 + \ldots + 2n = n(n + 1)$

    c) $1 + 4 + 7 + \ldots + (3n - 2) = \dfrac{n}{2}(3n - 1)$

    d) $1^2 + 2^2 + 3^2 + \ldots + n^2 = \dfrac{1}{6}n(n + 1)(n + 2)$

    e) $2 + 7 + 12 + \ldots + (5n - 3) = \dfrac{n}{2}(5n - 1)$

3. Find the distance between the following points:

    a) $P_1(3)$ and $P_2(7)$ b) $P_1(-3)$ and $P_2(7)$

    c) $P_1(12)$ and $P_2(4)$ d) $P_1(4)$ and $P_2(-3)$

    e) $P_1(-7)$ and $P_2(-4)$ f) $P_1(-5)$ and $P_2(-17)$

4. Find the set of real numbers given by the following:

    a) $[18, 27] \cap [9, 23]$ b) $[6, 9[\cup]1, 5]$

    c) $]12, 18[\cup [7, 17]$ d) $[1, 5] \cup]3, 7] \cap [2, 4]$

    e) $]5, 7] \cap [3, 6] \cup [1, 4]$ f) $[2, 4]'$

5. Perform the following operations:

    a) $(3 + 5i) + (7 - 2i)$ b) $(-2 + i) - (3 - 4i)$

   c) $(3 - i)(2 + 3i)$ d) $(-3 + 2i)(2 - 5i)$
   e) $(7 - i) \div (3 - 2i)$ f) $(8 + 3i) \div (-2 + i)$.
6. Prove that the set of rational numbers is dense.

IV

FUNCTIONS

40. Definitions. In this chapter we shall discuss the mathematical use of the word **function**. The concept of function is very basic in mathematics. In general terms the word *function* is used to denote a specific association between the elements of two sets. Thus, in order to have a function we shall need two sets, which we shall order as a first and second set, and then a rule that pairs a member of the second set with each member of the first set.

Let X be the first set and Y be the second set.

Definition. *A **function** f is an assignment by some rule of a unique (one and only one) element of Y to each element of X.*

Definition. *The set X is called **domain** of f and each element x ∈ X is called an **argument** of f.*

Definition. *The set Y is called the **range** of f and each element y ∈ Y is called a **value** of f.*

Since for each element $x \in X$ a unique element of $y \in Y$ is determined, we are forming a set of ordered pairs (x, y) in which to each x there is associated only one y. Thus a function f from X to Y is this set of ordered pairs and is a subset of the Cartesian product of X and Y. (See Section 24.)

The rule which assigns an element y to a specified argument x may take many forms. The most frequent form is that of an equation. The rule can also be a simple listing of the ordered pairs, or a graph, or a statement.

The notation for a function also takes many forms. The most frequent form in advanced mathematics is $f: X \rightarrow Y$, which is read "the function f from X to Y." If we want to emphasize the elements of the sets, we write $f: x \rightarrow y$, which is read "f takes x into y." The elements of the range are denoted by $f(x)$, which is read "the value of

f at x."* This notation permits the simple indication of the value of
the function for a specified x; e.g., $f(2)$ means the value (y) at $x = 2$.

Since a function from X to Y is a set of ordered pairs (x, y) with
$x \in X$ and $y \in Y$ we may use set notation to describe a function:

$$\{(x, y) | y = f(x)\}$$

In our discussion of a function we have called the elements of the
domain, $x \in X$, arguments of the function. Whenever the set X has
more than one element, the elements are also variables and they
may be arbitrarily selected. Consequently, the elements of the domain
are also called **independent variables** of the function. On the other
hand, the elements of the range, $y \in Y$, depend upon the choice of x
for their values. Thus, these elements are also called the **dependent
variables.**

We have been using the letter f to represent a function. Other
letters can be used, of course, and when more than one function is
discussed at a time, we frequently use letters like g, h, ϕ, etc.

41. Examples.   Due to the importance of the concept of function
we shall devote this section to a discussion of examples of functions.

Example 1.   Let the function f be given by the equation

$$y = 2x^2 - 8x + 1$$

and let the domain be the set of real numbers. We can now tabulate
as many ordered pairs as we please by giving values to x and calculating the corresponding values of y. Some of the ordered pairs are
$(0, 1), (1, -5), (-1, 11), (2, -7), (3, -5)$.

Example 2.   Consider a table of values

x	1	2	3	4
y	5	7	9	11

This table of values relates each value x to a unique value y and thus
establishes the set of ordered pairs $(1, 5), (2, 7), (3, 9), (4, 11)$. The
table defines a function whose domain is $X = \{1, 2, 3, 4\}$ and whose
range is $Y = \{5, 7, 9, 11\}$.

---

*The notation $f(x)$ was also used for many years to denote the function itself
and was read "f of x." This emphasized that we were discussing a function of
a single argument x. The notation is credited to Dirichlet (1805–1859), who
used it in the dual meaning. The reader will find this dual use in many books.

Example 3.   The area of a circle of radius r is given by the formula $A = \pi r^2$. The number π is a fixed value, a constant; thus the area A depends upon r for its value, and we can write $A(r) = \pi r^2$. Since we are considering a circle, the radius is a nonnegative number, $r \geq 0$. The formula defines a function whose domain is the set of nonnegative real numbers and whose range is also the set of nonnegative real numbers.

Example 4.   Consider all the cities in the state of Ohio. Each city is marked on a map of Ohio. The rule that associates a name with a position on the map is the function that takes $X = \{$names of cities$\}$ to $Y = \{$map positions$\}$.

Example 5.   The multiplicative-inverse axiom for real numbers says that there is a unique inverse $1/x$ for each $x \in R$ except $x = 0$. This can be expressed as the function $f: x \to 1/x$, whose domain and range is the set of real numbers R providing $x \neq 0$.

Example 6.   Consider the function f defined by

$$y = x + 2 \quad \text{if} \quad 0 \leq x \leq 3$$
$$y = 5 \quad \text{if} \quad x > 3$$

In this case two equations are required to define the function, depending upon the interval in which the elements of the domain are located. The domain is all nonnegative real numbers. The calculation of the range becomes an important operation and, as a general problem, will be discussed throughout the book. In this example we can calculate the following set of ordered pairs:

x	0	1	2	3	4	5	6
y	2	3	4	5	5	5	5

and notice that the values of y seem to fall between 2 and 5. The range is in fact $[2, 5]$. One method for verifying the range is to "graph the function," a concept we shall discuss in the next sections. This example also illustrates a special function. For all values of $x > 3$, the values of the function become the same fixed number, 5. Such a function is called a **constant function** and is written $y = c$ or $f: x \to c$ where it is assumed that c is a constant.

42. Coordinate System.   In order to give a pictorial representation of functions, mathematicians have devised a coordinate system that describes the position of a point in a plane. To define this coordinate system let us consider two number scales similar to that pictured in Figure 5, and let them intersect each other at right angles at the point O. We have now formed a **rectangular coordinate system.** The two lines are called the **axes,** denoted by OX and OY. See Figure 7. Let P be any point in the plane OX and OY.

The horizontal coordinate of P is the directed perpendicular distance, x, from OY to P. It is said to be *positive* when P is to the right of OY and *negative* when P is to the left of OY. The horizontal coordinate is called the **abscissa.**

The vertical coordinate of P is the directed perpendicular distance, y, from OX to P. It is said to be *positive* when P is above OX and *negative* when P is below OX. The vertical coordinate is called the **ordinate.**

The abscissa and ordinate together are called the **coordinates** of P and are denoted by (x, y). The point O is called the **origin,** and OP is called the **radius vector, r;** r is zero if P coincides with O; otherwise it is a positive quantity.

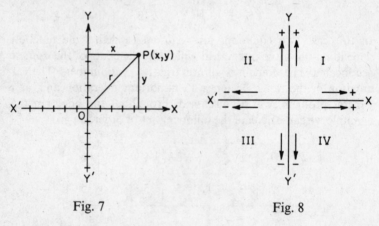

Fig. 7   Fig. 8

The coordinate axes divide the plane into four **quadrants** numbered counterclockwise I, II, III, and IV. The point P may be located on this plane simply by knowing the values of P's coordinates x and y. We see that in the first quadrant both x and y are positive; in the second quadrant x is negative, y positive; in the third quadrant $(-, -)$; and in the fourth quadrant $(+, -)$; see Figure 8.

To **plot** a point, $P(x, y)$, is to locate its position from the values of x and y.

Illustration. Plot the points $P_1(2,1), P_2(-3,2), P_3(-2, -4), P_4(3, -3)$.

Solution.   See Figure 9.

The coordinate system we have defined is usually called a **Cartesian coordinate system.**

43. Graph of a Function; Locus.   We can now give a pictorial representation of a function. Such a representation is called a **graph** or **locus** of a function. It is obtained by plotting arbitrarily assigned values of the independent variable as abscissas and the corresponding computed values of the function as ordinates. The first step is to obtain a table of values of the function for chosen values of the independent variable. Next, these pairs of values are plotted as points. The points are then joined by a smooth curve and we have a picture or graph of the function. In order to facilitate the plotting, graph paper is usually used.

Illustration.   Plot the graph of the function

$$y = 3x - 4$$

Solution.   Calculate the table of values.

x	-1	0	1	2	3
y	-7	-4	-1	2	5

Plot the corresponding points and join them with a curve; see Figure 10.

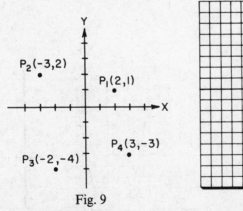

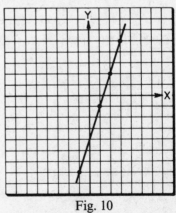

Fig. 9   Fig. 10

As many points as desired can be plotted simply by assigning more values to x and calculating the corresponding values of y from the function.

The term *locus* is important in mathematics, and although we have linked it with the term *graph*, it merits a more rigorous definition.

Definition. *The* **locus** *of an equation is the totality of all points, and only those points, whose coordinates satsify the equation.*

We must remember that (1) every point whose coordinates satisfy the equation lies on the locus, and that (2) the coordinates of every point on the locus satisfy the equation.

The locus is also thought of as the *path of a point,* $P(x, y)$, *which moves according to a specified law.*

Let us now return to the term *graph* and give a more rigorous definition.

Definition. *The* **graph** *of the equation* y = f(x) *is the totality or aggregate of all points whose coordinates* (x,y) *satisfy the relation* y = f(x).

44. Zero of a Function; Intercepts. If a function is equal to 0 for special values of the independent variable, the values are called **zeros** of the function. To explain further, let $y = f(x)$; then the zeros are those values of x for which $y = 0$. If we draw the graph of $y = f(x)$, the zeros are the x values or the abscissas of the point where the graph crosses or touches the X-axis.

Illustration. Plot the graph of the function y given by

$$y = x^2 + x - 6$$

and find the zeros.
Solution.

x	y
0	-6
1	-4
2	0
3	6
-1	-6
-2	-4
-3	0
-4	6

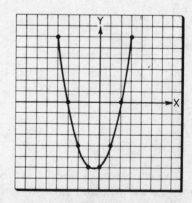

The zeros are $x = 2$, $x = -3$. 　　　　　Fig. 11

In geometry the value of x for which the curve crosses the X-axis is called the **x-intercept.** Similarly, the value of y for which the curve crosses the Y-axis is called the **y-intercept.** To obtain these intercepts we set y equal to 0 and solve for x, then set x equal to 0 and solve for y.

In the above illustration the x-intercepts are the zeros of the function, $x = 2$, $x = -3$. The y-intercept is $y = -6$.

45. Graph of a Function; Image.   In the previous sections we have been discussing the graphs of functions which have been defined by equations. We have plotted a finite number of points on a rectangular coordinate system and joined these points by a smooth curve, and we assumed that the points on the curve which were between two calculated points also satisfied the equation. Since it is possible to define a function without using an equation, let us consider the concept of a graph for such functions.

Illustration.   Plot the graph of the function y, if

$$y = \begin{cases} -1 \text{ if } x \text{ is an odd integer} \\ 0 \text{ if } x = 0 \\ 1 \text{ if } x \text{ is an even integer} \end{cases}$$

Solution.   We first note that the domain is limited to the set of integers and that the values of the function are limited to -1, 0, and 1. The graph, which is a set of discrete points, is shown in Figure 12.

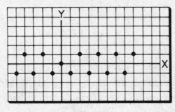

Fig. 12

A function f is often thought of as a rule which maps the set X into the set Y, and the graph is also called the **image** of the set X under the mapping of f. This is especially true when we consider a function that takes a set of ordered pairs into another set of ordered pairs.*

46. Special Functions.   In this section we shall discuss three special functions.

---

*This concept is studied extensively when we are dealing with complex variables. See R. V. Churchill, *Complex Variables and Applications,* New York: McGraw-Hill Book Company, Inc., 1960, pp. 19-22 and Chapter 4.

The Absolute Value Function. The absolute value of a number was defined in Section 34. Consider now the function f defined by

$$y = |x| = \begin{cases} x \text{ if } x > 0 \\ 0 \text{ if } x = 0 \\ -x \text{ if } x < 0 \end{cases}$$

where the domain is the set of real numbers. Calculate some of the ordered pairs,

x	-2	-1	0	1	2	3	7
y	2	1	0	1	2	3	7

Since the values of the function are all nonnegative, we see that the range is the set of nonnegative members of the real numbers, R. The graph is shown in Figure 13.

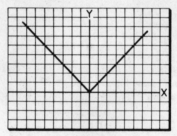

Fig. 13

The Identity Function. The function defined by the equation $y = x$ is called the **identity function.** It is easily seen that the values of the function are exactly the same as the values of the argument. The graph is a straight line through the origin at a $45°$ angle to either of the axes.

The Greatest Integer Function. We introduce a new symbol:

$[x]$ is read "the greatest integer not greater than x."

Examples:

a) $[2] = 2$ b) $[\frac{3}{2}] = 1$ c) $[\frac{132}{5}] = 26$ d) $[-\frac{3}{2}] = -2$

The function f whose values are given by

$$f(x) = [x]$$

and whose domain is the set of real numbers is called the **greatest integer function.** We notice that the values of the function remain the same integer as x takes on the values between two consecutive integers. The graph is shown in Figure 14.

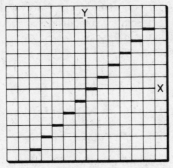

Fig. 14

47. The Algebra of Functions.   We can apply the operations of algebra to the concepts of functions. Let us consider the elementary operations and define their application as follows:

Definitions:

1. *In the* **addition,** f + g, *of two functions, the values of the functions are added,*

$$f : (x, y) + g : (x, z) = f + g : (x, y + z)$$

2. *In the* **subtraction,** f − g, *of two functions, the values of the functions are subtracted in the stated order,*

$$f : (x, y) - g : (x, z) = f - g : (x, y - z)$$

3. *In the* **multiplication,** fg, *of two functions, the values of the functions are multiplied,*

$$f : (x, y) \times g : (x, z) = (fg) : (x, yz)$$

4. *In the* **division,** f/g, *of two functions, the values of the functions are divided in the stated order,*

$$\frac{f : (x, y)}{g : (x, z)} = \left(\frac{f}{g}\right) : \left(x, \frac{y}{z}\right), z \neq 0$$

In the above definitions we shall let the domain of f be denoted by d_f and that of g by d_g; then the domain of the sum, difference, product, and quotient of f and g is the intersection of the sets d_f and d_g; i.e., $d_f \cap d_g$. In the case of the quotient, those values of $x \in d_f \cap d_g$

for which $z = g(x) = 0$ are excluded since the division by zero is not permitted.

Illustrations.

1) Given the functions f defined by $y = x$ with domain $d_f = R$ (set of real numbers) and g defined by $z = \sqrt{1 - x^2}$ with d_g equal to the interval $-1 \le x \le 1$. Find the sum, difference, product, and quotient of f and g and the domain for each.

Solution. By definition

$$f + g = x + \sqrt{1 - x^2}, \quad d_{f+g} \text{ is } -1 \le x \le 1$$
$$f - g = x - \sqrt{1 - x^2}, \quad d_{f-g} \text{ is } -1 \le x \le 1$$
$$fg = x\sqrt{1 - x^2}, \quad d_{fg} \text{ is } -1 \le x \le 1$$
$$\frac{f}{g} = \frac{x}{\sqrt{1 - x^2}}, \quad d_{f/g} \text{ is } -1 < x < 1$$

Note that $d_{f/g}$ is the open interval $]-1, 1[$ since $g(-1) = g(1) = 0$.

2) Find $f + g, f - g, fg$, and f/g if f is defined by $y = 1 + x$ and g is defined by $y = 1 - x^2$ and $d_f = d_g = R$.

Solution.

$$f + g = (1 + x) + (1 - x^2) = 2 + x - x^2, \quad d_{f+g} = R$$
$$f - g = 1 + x - (1 - x^2) = x + x^2, \quad d_{f-g} = R$$
$$fg = (1 + x)(1 - x^2) = 1 + x - x^2 - x^3, \quad d_{fg} = R$$
$$\frac{f}{g} = \frac{1 + x}{1 - x^2}, \quad d_{f/g} = R, \quad x \ne \pm 1$$

Let us consider three sets X, Y, and Z and two functions f and g defined such that $f : X \to Y$ and $g : Y \to Z$. Here the set X is made to correspond to a set Y by means of the function f; after which the set Y is made to correspond to the set Z by means of the function g; thus X is mapped into the set Z. In terms of the elements we can define the values $z = g(y)$ and $y = f(x)$ and then say that $z = g(f(x))$. We give the concept a formal definition.

Definition. *If* f : X → Y *and* g : Y → Z, *then* g(f) : X → Z *is called the* **composite** *of* g *and* f.

There are many common notations for the composite of two functions, such as $g \circ f : (x, z)$, $g[f]$, with its value at x by $(g \circ f)(x)$, $g[f](x)$, or simply $z = g(f(x))$. Let f be a function with domain d_f and range r_f and g be a function with domain d_g and range r_g. Let A be the subset of d_f whose elements x have values that belong to $r_f \cap d_g$. The function g assigns to each element of $r_f \cap d_g$ some element of r_g. Thus there exists a function h with domain A and for which the values $h(x) = g(f(x))$ for each $x \in A$ form a range which is

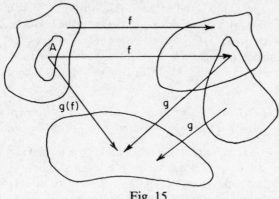

Fig. 15

a subset of r_g. See Figure 15.

Illustration.   If f is defined by $y = x + 3$ and g is defined by $z = 2y^2 - 3$, find the composite $g {\circ} f$.
Solution.

$$g {\circ} f \text{ has values } z = g(f(x)) = 2(x + 3)^2 - 3$$
$$= 2(x^2 + 6x + 9) - 3$$
$$= 2x^2 + 12x + 15$$

Suppose we are given two functions f and g whose values are defined by $f(x)$ and $g(x)$. We can now form two composites $f {\circ} g$ and $g {\circ} f$ which are not necessarily the same.

Illustration.   Let the functions f and g be defined by

$$f(x) = x^2 - 3 \quad \text{and} \quad g(x) = 3x - 5$$

Find the composites $g {\circ} f$ and $f {\circ} g$.
Solution.
To find $g {\circ} f$ write

$$y = f(x) = x^2 - 3 \quad \text{and} \quad z = g(y) = 3y - 5$$

then $g {\circ} f$ has values

$$z = g(f(x)) = 3(x^2 - 3) - 5 = 3x^2 - 14$$

To find $f {\circ} g$ write

$$y = 3x - 5 \quad \text{and} \quad z = x^2 - 3$$

then $f {\circ} g$ has values

$$z = f(g(x)) = (3x - 5)^2 - 3 = 9x^2 - 30x + 22$$

48. Relations.   In defining a function we stated that a *unique* element of Y was associated with each element of X. It is, of course, possible to specify a correspondence between the two sets X and Y such that for some x there are two or more values of y. Such a correspondence would form a set of ordered pairs in which the second element is not necessarily unique; we shall give this concept a name.

Definition. *The assignment of an element of* Y *to each element of* X *to form a set of ordered pairs* (x, y) *is called a* **relation.**

A relation has a domain (the set of all first elements) and a range (the set of all second elements), just as does a function. In fact, functions are special cases of relations, namely, those relations for which the second element of the ordered pair is unique.*

Let us consider some examples of relations that are not functions. *Examples:*

a) $\{(x, y)|y^2 + x^2 = 1\}$
b) $\{x, y)|y^2 = x\}$
c) $\{(x, y)|y^4 = x^2 + 2x + 1\}$

We can plot the graphs of relations in the same manner that we graph functions.

Illustration. Plot the graph of the relation defined by $y^2 = 4x$.
Solution. Calculate a few ordered pairs by assigning values to x.

x	0	1	2	4
y	0	±2	$\pm\sqrt{8}$	±4

Note that we have two values of y for each x (except $x = 0$). If we plot these ordered pairs in a rectangular coordinate system and join them with a smooth curve, we obtain the graph shown in Figure 16.

49. Inverse Function. A function is an association that creates a set of ordered pairs (x, y) such that for each x there is a unique y. However, different values of x can have the same y. If the function f is such that each x has a different and unique y so that no two pairs have the same second element, then there is a one-to-one corres-

---

*Classical mathematical literature treats the above distinction by referring to functions as "single-valued" or "multiple-valued."

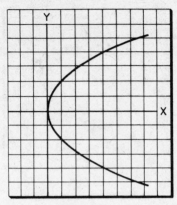

Fig. 16

pondence between the set X and the set Y and we can define a very important concept.

Definition. *If the function* f *is such that no two of its ordered pairs have the same second element, then there exists a function* f^{-1}, *called the* **inverse function** *of* f, *which is obtained from* f *by interchanging the first and second elements in each ordered pair.*

We see that the range of f is the domain of f^{-1} and the domain of f is the range of f^{-1}. Some authors prefer to use the domain and range in the definition of the inverse function. If the function f is defined by an equation, the inverse function is found by interchanging x and y in the original equation and solving for y to obtain an explicit equation that defines f^{-1}.

Illustration. If f is defined by the equation

$$f(x) = y = 3x + 1$$

over the domain $0 \le x \le 5$, find the inverse of f.
Solution: Interchange x and y in the equation:

$$x = 3y + 1$$

and solve for y to obtain

$$y = \frac{x-1}{3} = f^{-1}(x)$$

The range of f is $1 \le y \le 16$, which is the domain of f^{-1}, and the range of f^{-1} is $0 \le y \le 5$. The graphs of f and f^{-1} are shown in Figure 17, p. 82.

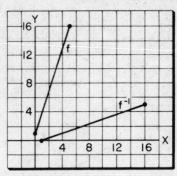

Fig. 17

If the function f does not give a one-to-one correspondence between X and Y, it may still be possible to find an inverse; however, the inverse will not be a function, it will be a relation.

Example: If f is given by $y = x^2$, then the inverse is $y^2 = x$ or $y = \pm\sqrt{x}$, which is a relation. Note that the range of f is the set of positive real numbers (for $d_f = R$) and that this range is the domain of the inverse, so that the range of the inverse is the set of real numbers.

Summary of Symbols	
f, function d_f, domain of function f $f(x)$, value of function f at x	f^{-1}, inverse function r_f, range of function f
$f: X \to Y$, the function f from the set X to the set Y $f: x \to y$, f takes the element x into the element y $[x]$, the largest integer not greater than x $g[f]$ and $g \circ f$, the composite of g and f	

50. Exercise V.

1. Which of the following sets describe a function and which describe a relation?

   a) $\{(1, 2), (3, 4), (5, 6)\}$ b) $\{(1, 2), (2, 3), (1, 3), (3, 5)\}$

   c) $\{(1, 2), (2, 2), (3, 2)\}$ d) $\{(3, 2), (2, 1), (1, 0)\}$

   e) $\{(x, y) | y = 2x - 5\}$ f) $\{(x, y) | y + x = 1\}$

   g) $\{(x, y) | y^2 = 2x + 1\}$ h) $\{(t, v) | v = t^2 - 16\}$

   i) $\{(x, y) | x^2 + y^2 = 9\}$ j) $\{(r, s) | r = s^2 + 2\}$

   k) $\{(x, y) | x = y^{-1}\}$ l) $\{(x, y) | x + y = a^2\}$

m) $\{(x, y)|y = 10^x\}$ n) $\left\{(x, y)\Big|y = \left[\dfrac{x}{a}\right]\right\}$

2. Find the values, $f(x)$, for $x = 0, 1, 2, 3, -1, -2, -3$ in each of the following:
   a) $f(x) = x^2 + 1$ b) $f(x) = x + 3$
   c) $f(x) = -(x - 3)$ d) $f(x) = \sqrt{x} + 1$
   e) $f(x) = \dfrac{1}{x - 1}$ f) $f(x) = \pi x^2$

3. Plot the graphs of the following relations over the domain $-5 \le x \le 5$:
   a) $y = 2x - 3$ b) $x - y = 7$
   c) $y = x^2 + 1$ d) $y = 1/x$
   e) $y = \sqrt{5} - x$ f) $x^2 + y^2 = 4$
   g) $y = |x|$ h) $y = [x]$
   i) $y = \begin{cases} 3 \text{ if } x < 0 \\ 2 \text{ if } x = 0 \\ 1 \text{ if } x > 0 \end{cases}$ j) $y = \begin{cases} x + 5 \text{ if } x < 0 \\ 0 \text{ if } x = 0 \\ x - 5 \text{ if } x > 0 \end{cases}$
   k) $\{(-5, 4), (-5, -4), (-4, 3), (-4, -3), (-3, 2),$
      $(-3, -2), (-2, 1), (-2, -1), (-1, 0)\}$
   l) $|y| + |x| = 5$

4. Find the values of $f + g$, $f - g$, fg, and f/g if f and g are defined by the following equations:
   a) $f(x) = x^2 - 3$, $g(x) = x$
   b) $f(x) = \dfrac{1}{x + 1}$, $g(x) = x - 1$
   c) $f(x) = \dfrac{1}{(x + 1)^2}$, $g(x) = x + 1$
   d) $f(x) = x + 1$, $g(x) = x - 1$
   e) $f(x) = 1^x$, $g(x) = x + 1$
   f) $f(x) = a$, $g(x) = b$

5. Find the composite of g and f:
   a) $z = g(y) = y^2 + y$, $y = f(x) = x - 1$
   b) $z = g(y) = y^{-2}$, $y = f(x) = x^{-2}$
   c) $z = g(y) = |y + 3|$, $y = f(x) = x - 3$

6. Find f^{-1} if f is defined by the following equations:
   a) $y = 3x - 2$ b) $y = x + 5$
   c) $y = x$ d) $y = x + k$
   e) $y = cx$ f) $y = 3 - x$
   g) $ax + by = c$ h) $y = mx + b$

7. Let the domain of definition of f be the set of integers I. Find the range of f and plot the graph if f is given by the following equations:

   a) $y = x$ b) $y = x^2$

8. Given the function f defined by $f(x) = x + 1$, find $f(a)$, $f(x + h)$, $f(x - h)$, $f(x + h) - f(x)$, and $[f(x + h) - f(x)] \div h$.

9. If we limit our discussion to the set of real numbers R, find the domains for each of the following functions:

   a) $\{(x, y)| y = x^{-1}\}$
   b) $\{(x, y)| y = \sqrt{x}\}$
   c) $\{(x, y)| y = (x - 1)^{-1} + (x - 2)^{-1}\}$

10. Let f be defined by $y = x^2$ with $d_f = R$ and g be defined by $z = x^3$ with $d_g = R$. Find the domains of $f + g$, $f - g$, fg, and f/g.

V

LINEAR ALGEBRA

51. Introduction. In this chapter we shall discuss the algebra of linear expressions. We defined the linear polynomial in Section 20 and the reader is already familiar with the simple open sentence

$$ax + b = 0$$

where x is a variable and a and b are constants. The procedures for finding the truth set for a linear equation were discussed in Section 20. Let us consider a few typical problems which lead to linear equations.

If the equation is given in fractional form, multiply through by the lowest common denominator (L.C.D.).

Illustration. Solve

$$\frac{x}{x-2} - \frac{2x}{x+2} = 2 - \frac{3x^2}{x^2 - 4}.$$

Solution. Clear the equation of fractions by multiplying by the L.C.D. $(x-2)(x+2)$ and simplify:

$$x(x+2) - 2x(x-2) = 2(x^2 - 4) - 3x^2$$
$$x^2 + 2x - 2x^2 + 4x = 2x^2 - 8 - 3x^2$$
$$-x^2 + 6x = -x^2 - 8$$
$$x = -\tfrac{4}{3}$$

Check.

$$\frac{-\tfrac{4}{3}}{(-\tfrac{4}{3}) - 2} - \frac{2(-\tfrac{4}{3})}{(-\tfrac{4}{3}) + 2} = 2 - \frac{3(-\tfrac{4}{3})^2}{(-\tfrac{4}{3})^2 - 4}$$

$$\tfrac{22}{5} = \tfrac{22}{5}$$

In solving physical problems (or "word problems") the essential procedure is to convert the words in the problem into the symbols and equations of algebra. A thorough reading and understanding of the problem is the first step; the algebra itself is usually not very difficult.

Illustration. A man sold 2 acres less than $\frac{3}{4}$ of his farm. He then had 4 acres less than $\frac{1}{2}$ of it left. How many acres were on his original farm?
Solution. Let x = number of acres in original farm. Using the principle that the sum of the parts equals the whole, we write

$$\tfrac{3}{4}x - 2 + \tfrac{1}{2}x - 4 = x$$

Solving for x we get

$$(\tfrac{3}{4} + \tfrac{1}{2} - 1)x = 6$$
$$x = 24 \text{ acres}$$

The relevant physical laws or assumptions must be known before word problems can be translated into algebraic equations. We must, for example, know that distance traveled equals velocity times time; that the area of a rectangle is the length times the width; etc. Many of these concepts will be explained in this book; others the student will already know; still others may be found in advanced books to which we may refer.

Illustration. A and B start from the same place and travel in the same direction with B starting $\frac{1}{2}$ hour after A. If A travels at a rate of 50 miles per hour, and B, 60 miles per hour, how far will they have traveled when B overtakes A?
Solution. Let t = the number of hours B has traveled. Then $t + \frac{1}{2}$ = number of hours A had traveled. The distance d is the average speed times the time. Thus:

$$\begin{aligned} \text{A travels} \qquad & d = 50(t + \tfrac{1}{2}) \\ \text{B travels} \qquad & d = 60t \end{aligned}$$

Now B overtakes A when they have both traveled the same distance, so that:

$$60t = 50(t + \tfrac{1}{2})$$

Solving for t, we have

$$10t = 25 \qquad \text{or} \qquad t = 2.5 \text{ hours}$$

Check. As a check let us compute the distance traveled.

$$\begin{aligned} \text{For A} \qquad & d = 50(t + \tfrac{1}{2}) = 50(3) = 150 \text{ miles} \\ \text{For B} \qquad & d = 60t = 60(2.5) = 150 \text{ miles} \end{aligned}$$

A very important principle in the solution of equations is the following: *If a product of two or more factors equals zero, one or more of its factors equals zero.*

Illustration. In solving the equation

$$(3x - 9)(x + 2) = 0$$

we set each factor equal to zero.

$$\overline{3x - 9 = 0 \mid x + 2 = 0}$$
$$x = 3 \mid \quad x = -2$$

Check.

$$(3 \cdot 3 - 9)(3 + 2) = (0)(5) = 0$$
$$[(3)(-2) - 9][-2 + 2] = (-15)(0) = 0$$

We shall concern ourselves not only with linear equations but also with functions defined by linear equations, their graphical representations, linear inequalities, systems of linear equations, and some applications.

52. The Linear Function. Consider the function f defined by the equation

$$y = mx + b$$

over the domain of real numbers R and limit the constants m and b to be elements of R. Since x occurs to no higher degree than the first, this function is called a **linear function.** For given values of the constants m and b, the set of ordered pairs (x, y) can be calculated and we can plot the graph of f on a rectangular coordinate system. If we proceed as we have previously, by joining a finite number of points with a smooth curve, the curve will appear to be a straight line. The fact that the graph is a straight line must be proved (see Section 54); however, for the present we shall assume the truth of the theorem.

Theorem. *The graph of a linear function is a straight line.*

If we limit the domain of definition of the function defined by $y = mx + b$ to a closed interval, say $[c, d]$, then the graph of the function is a line segment with end-points $P_1(c, y_1)$ and $P_2(d, y_2)$ where

$$y_1 = mc + b \quad \text{and} \quad y_2 = md + b$$

The range of this function is the interval $[y_1, y_2]$.

Illustration. Plot the graph of the functions defined by the following equations.
a) $y = 3x - 5$, $\quad d_f = R$
b) $y = 3 - x$, $\quad d_f = [-3, 2]$
Solution.
a) The graph is shown in Figure 18, p. 88, and $r_f = R$.
b) The graph is shown in Figure 19, p. 88. The range is $[1, 6]$.

Before discussing additional characteristics of a linear function,

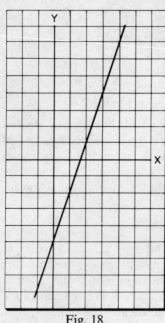

Fig. 18

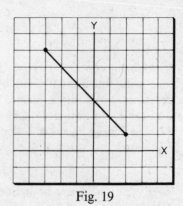

Fig. 19

let us consider some of the geometric properties of a straight line.

53. Slope of a Line.   Let us choose two points $P_1(x_1, y_1)$ and $P_2(x_2, y_2)$ which lie on a given line.

Definition.* *The ratio*

$$m = \frac{y_2 - y_1}{x_2 - x_1}$$

is called the **slope** *of the line through* P_1 *and* P_2.

Illustration.   Find the slopes of the lines through the following points.
a) $P_1(3, 5)$ and $P_2(7, 8)$
b) $P_1(-1, 3)$ and $P_2(2, -1)$
c) $P_1(2, -3)$ and $P_2(1, -5)$
Solution.   By definition we have

a) $m = \dfrac{8 - 5}{7 - 3} = \dfrac{3}{4}$

b) $m = \dfrac{-1 - 3}{2 + 1} = -\dfrac{4}{3}$

c) $m = \dfrac{-5 + 3}{1 - 2} = \dfrac{-2}{-1} = 2$

---

*For those who have studied trigonometry, the slope of a line can be defined as the tangent of the angle of inclination with the X-axis.

Note that it is immaterial which point we pick as P_1 or P_2 as long as we are consistent in formulating the differences.

Example: Interchange P_1 and P_2 in Illustration (*a*) above; then

$$m = \frac{5 - 8}{3 - 7} = \frac{-3}{-4} = \frac{3}{4}$$

A line parallel to the X-axis has $y_2 = y_1$; therefore $y_2 - y_1 = 0$ and the slope is 0. A line parallel to the Y-axis has $x_2 = x_1$ and $x_2 - x_1 = 0$; this would lead to a division by 0 in finding the slope. Since division by 0 is not permitted, the slope of the line parallel to the Y-axis is undefined. The graphical representation of the identity function defined by $y = x$ is a straight line through the origin inclined at an agle of $45°$ to the positive X-axis. Two points on this line would have $y_1 = x_1$ and $y_2 = x_2$, so that

$$m = \frac{y_2 - y_1}{x_2 - x_1} = \frac{x_2 - x_1}{x_2 - x_1} = 1$$

54. Forms Defining a Straight Line. There are many different ways in which we can write the equation of a straight line. We shall now list some of the more important ones.

I. *Parallel to a coordinate axis.* If a straight line is parallel to the Y-axis, its equation is

(1a) $$x = k$$

If a straight line is parallel to the X-axis, its equation is

(1b) $$y = k$$

These forms are clear if we consider the equations to read that $x = k$ (or $y = k$) for all values of y (or x).

Illustration. Plot the line $x = 3$.
Solution. We let $x = 3$ for all values of y, thus:

y	0	1	2	3	4	etc.
x	3	3	3	3	3	

and obtain the plot of Figure 20, p. 90.

II. *Point-slope form.* The equation of a straight line passing through a given point $P_1(x_1, y_1)$ having a given slope m is

(2) $$y - y_1 = \mathbf{m}(x - x_1)$$

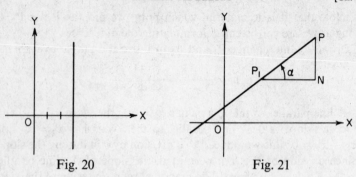

Fig. 20   Fig. 21

Proof. Construct Figure 21, where $P(x, y)$ is any point on the line through $P_1(x_1, y_1)$. Note that

$$\frac{NP}{P_1N} = \frac{y - y_1}{x - x_1} = m$$

or
$$y - y_1 = m(x - x_1)$$

Illustration. Find the equation of the line with a slope of -2 and passing through $(3, 1)$.

Solution.

$$y - 1 = -2(x - 3)$$
$$y = -2x + 7$$

III. *Two-point form.* The equation of a straight line through two given points $P_1(x_1, y_1)$ and $P_2(x_2, y_2)$ is

(3) $$y - y_1 = \frac{y_2 - y_1}{x_2 - x_1}(x - x_1)$$

This form is obtained simply by substituting for the slope in equation (2), since by Section 53

$$m = \frac{y_2 - y_1}{x_2 - x_1}$$

If $x_2 = x_1$, the slope is undefined as $x_2 - x_1$ becomes 0. However, formula (3) may be written in the form

(3a) $$(y - y_1)(x_2 - x_1) = (y_2 - y_1)(x - x_1)$$

Then, if $x_2 = x_1$, we have

$$0 = (y_2 - y_1)(x - x_1) \quad \text{or} \quad x = x_1$$

which is a line parallel to the Y-axis.

Illustration. Find the equation of a line through the points $(-3, 2)$ and $(2, -5)$.

Solution. We have

$$y_2 - y_1 = -5 - 2 = -7$$
$$x_2 - x_1 = 2 - (-3) = 5$$

Substituting into formula (3) we obtain

$$y - 2 = -\tfrac{7}{5}(x + 3)$$

or

$$5y - 10 = -7x - 21$$
$$5y + 7x + 11 = 0$$

Check.

$$5(2) + 7(-3) + 11 = 0 \qquad 5(-5) + 7(2) + 11 = 0$$

Note that it is immaterial which point we pick as P_1 or P_2 as long as we are consistent. Reverse the order of the points in the illustration and solve the problem.

IV. *Slope-intercept form.* The equation of a straight line whose slope is m and whose y-intercept (see Section 44) is b is

$$(4) \qquad\qquad y = mx + b$$

The point where the line crosses the Y-axis is $P_1(0, b)$. Substituting this point into formula (2) we obtain

$$y - b = m(x - 0)$$

or

$$y = mx + b$$

This is the usual linear function form; see Section 52.

Illustration. Find the equation of a line with slope equal to -2 and crossing the Y-axis at $y = 3$.

Solution. We have $m = -2$ and $b = 3$; thus

$$y = -2x + 3$$

or

$$y + 2x = 3$$

Check. $P_1(0, 3)$: $\qquad 3 + 2(0) = 3$

V. *The general form.* The general form of the equation of a straight line is given by

$$(5) \qquad\qquad Ax + By + C = 0$$

We note some properties:

    a) If $A = 0$, the line is parallel to the X-axis.

    b) If $B = 0$, the line is parallel to the Y-axis.

    c) If we solve for y, we obtain the slope-intercept form and can rea off the slope and y-intercept.

$$y = -\frac{A}{B}x - \frac{C}{B}$$

Thus $\qquad\qquad m = -\dfrac{A}{B} \quad$ and $\quad b = -\dfrac{C}{B}$

Illustration. Find the slope and y-intercept of

$$3x - 2y + 6 = 0$$

Solution. We solve the equation for y:

$$2y = 3x + 6$$
or $\qquad\qquad\qquad y = \tfrac{3}{2}x + 3$

Then $\qquad\qquad m = \tfrac{3}{2} \quad$ and $\quad b = 3$

VI. *Intercept form.* The equation of a straight line with intercepts a and b is

(6) $\qquad\qquad\qquad\qquad \dfrac{x}{a} + \dfrac{y}{b} = 1$

Let us find the equation of a line through the points $P_1(a, 0)$ and $P_2(0, b)$. By formula (3):

$$y - 0 = \frac{b}{-a}(x - a)$$

Simplify to obtain $\qquad ay = -bx + ab$
or $\qquad\qquad\qquad\quad ay + bx = ab$

Divide by ab to obtain formula (6), which is called the *intercept form.* It is clear that this form fails if the line passes through the origin, $(0, 0)$. If we know the intercepts we can draw the line simply by joining the points $(a, 0)$ and $(0, b)$.

Illustration. Find the intercepts and draw the line for the equation

$$3x + 2y = 6$$

Solution. To obtain the intercepts let $x = 0$ and solve for y; then let $y = 0$ and solve for x.

$$x = 0, \quad 2y = 6 \quad \text{or} \quad y = 3 = b$$
$$y = 0, \quad 3x = 6 \quad \text{or} \quad x = 2 = a$$

Locate the points $(2, 0)$ and $(0, 3)$ and join them with a straight line.

We can now prove the theorem stated in Section 52.

Theorem. *An equation of a straight line is of the first degree in* x *and* y; *and conversely, the graph of any equation of the first degree in* x *and* y *is a straight line.*

Proof. Consider the straight line, shown in Figure 22, with two

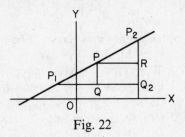

Fig. 22

fixed points $P_1(x_1, y_1)$ and $P_2(x_2, y_2)$. Choose a representative point P on the line with the variable coordinates (x, y). Through the points P, P_1, and P_2 draw lines parallel to the axes, forming the three similar triangles and intersections $Q(x, y_1)$, $Q_2(x_2, y_1)$, and $R(x_2, y)$, shown in the figure. Since the triangles are similar, the corresponding sides are proportional and we can write the ratios for $\triangle P_1QP$ and $\triangle P_1Q_2P_2$.

$$\frac{P_1Q}{P_1Q_2} = \frac{PQ}{P_2Q_2}$$

It was shown in Section 37 that the length of a line segment can be expressed in terms of the coordinates of its end-points. If we order the points consistently (e.g., right minus left) we may write

$$P_1Q = x - x_1 \qquad PQ = y - y_1$$
$$P_1Q_2 = x_2 - x_1 \qquad P_2Q_2 = y_2 - y_1$$

Substituting these values into the ratio equation, we obtain

$$\frac{x - x_1}{x_2 - x_1} = \frac{y - y_1}{y_2 - y_1}$$

Since the letters with subscripts are constants, their differences are constants so that we can write

$$x_2 - x_1 = a \qquad y_2 - y_1 = b$$

and the equation may be changed as follows:

$$a(y - y_1) = b(x - x_1)$$
$$ay - bx - ay_1 + bx_1 = 0$$
$$Ay + Bx + C = 0$$

which is a first-degree equation in x and y.

Converse. The general equation of the first degree in two variables is

$$Ax + By + C = 0$$

with the restriction that both A and B cannot be zero.

If $A = 0$ and $B \neq 0$, then $By + C = 0$ has the form of equation (1b) and the graph is a line parallel to the X-axis.

If $A \neq 0$ and $B = 0$, then $Ax + C = 0$ has the form of equation (1a) and the graph is a line parallel to the Y-axis.

If $A \neq 0$ and $B \neq 0$, then we can find two elements of the solution set, (x_1, y_1) and (x_2, y_2), to satisfy the equation:

$$Ax_1 + By_1 + C = 0$$
$$Ax_2 + By_2 + C = 0$$

This system can be solved for A and B in terms of x_1, y_1, x_2, y_2, and C. (See Section 56.) A substitution of the solution into the equation yields

$$(y_2 - y_1)x - (x_2 - x_1)y + (x_2y_1 - x_1y_2) = 0$$

which is the two-point form of the equation of a straight line. This completes the proof of the theorem.

55. Rate of Change of a Linear Function. When two variables are related through an equation, it is interesting to note what happens to one of the variables as the other variable is changed. To study this question let us introduce the mathematical notation, Δx, read "delta x"—Δx is *one quantity, not* Δ times x. This symbol is used when we are varying x and denotes the *increment* by which the variable is changed. Thus, if we change the value of x from x_1 to x_2, we can write

$$\Delta x = x_2 - x_1$$

Example: Let $y = 2x - 1$ and assign values to x as shown in the table (p. 95). We compute the corresponding values of y and the increments.

The general notation for changing x is to add an increment, Δx, to x and write $x + \Delta x$. If we let $y = f(x)$ and add an increment to x, we have a change in y and we write

$$y + \Delta y = f(x + \Delta x)$$

The change in y can be found by subtracting y from both sides of

x	Δx	y	Δy
1		1	
	1		2
2		3	
	1		2
3		5	
	2		4
5		9	
	4		8
9		17	

this equation to obtain

$$\Delta y = f(x + \Delta x) - y = f(x + \Delta x) - f(x)$$

We emphasize that when we write $f(x + \Delta x)$ we replace x by $x + \Delta x$ everywhere in the expression for $f(x)$.

Example: If $f(x) = 2x^2 - 3x$, then

$$f(x + \Delta x) = 2(x + \Delta x)^2 - 3(x + \Delta x)$$
$$= 2[x^2 + 2x\Delta x + (\Delta x)^2] - 3x - 3\Delta x$$
$$= 2x^2 + (4\Delta x - 3)x + 2(\Delta x)^2 - 3\Delta x$$

Definition.   *The **average rate of change** of a variable* y *with respect to a variable* x *within a given interval is the quotient* $\Delta y/\Delta x$.

Theorem.   *The rate of change of a linear function is constant and equal to the slope of its straight line.*

Proof.   A linear function can be defined by the equation

$$y = mx + b$$

By definition we have

$$\frac{\Delta y}{\Delta x} = \text{rate of change of } y \text{ with respect to } x$$

We proceed to find this quotient.

$$y + \Delta y = m(x + \Delta x) + b$$
$$\Delta y = mx + m\Delta x + b - mx - b$$
$$= m\Delta x$$

Therefore

$$\frac{\Delta y}{\Delta x} = m$$

which is a constant and the slope of the line represented by the linear equation.

Note that a *positive* rate means that the function is *increasing* as the independent variable increases; *negative* rate, *decreasing*. Draw two lines, one with a positive slope and one with a negative slope; investigate the behavior of y for each line as x increases.

56. Systems of Linear Equations. The reader is already familiar with the concept of finding a solution of an equation (see Section 20). The linear equation in one unknown has one and only one solution, which can be found by applying the axioms of equality and the operations given in Section 20.

Let us turn our attention to linear equations in more than one variable. The linear equation

$$Ax + By + C = 0$$

has an unlimited number of solutions. In fact, the solution set is composed of the infinite number of elements,

$$S = \{(x_1, y_1) | Ax_1 + By_1 + C = 0\}$$

Geometrically, every point on the straight line which represents this equation satisfies the equation.

Consider now two equations in the same two variables:

$$A_1 x + B_1 y + C_1 = 0$$
$$A_2 x + B_2 y + C_2 = 0$$

When these equations are considered at the same time, they are said to form a *system of simultaneous linear equations*. A *solution set* of this system is a pair of corresponding values of x and y which satisfy both of the equations. Three things can occur in this system. Since each of the equations has a straight line for its graph, we may discuss the possibilities in terms of the straight lines.

a) If the straight lines have only one point in common, the

coordinates of this point constitute a solution set of the system, and the equations are said to be **consistent**.

b) If the two lines are parallel, they have no point in common, there is no solution (the solution set is $\emptyset$), and the equations are said to be **inconsistent**.

c) If the two lines are coincident, all points are common, there is an infinite number of solutions, and the equations are said to be **dependent**.

The three cases can be checked by considering the slopes of the lines. (See also Section 57.) If the slopes are *equal*, the lines are either *parallel* or *coincident*. If the slopes are *not equal*, the system has one and only one solution.

Illustrations. Determine the nature of the following systems of linear equations:

1)
$$\begin{cases} 3x - 2y - 6 = 0 \\ 2x - y + 3 = 0 \end{cases}$$

Solution. $m_1 = \frac{3}{2}$, $m_2 = 2$, $\therefore$ Consistent

2)
$$\begin{cases} 2x + y = 4 \\ 4x + 2y = 8 \end{cases}$$

Solution.

$m_1 = -2$, $m_2 = -\frac{4}{2} = -2$, $\therefore$ Inconsistent or dependent

3)
$$\begin{cases} x = y - 5 \\ x = y + 7 \end{cases}$$

Solution. $m_1 = 1 = m_2$, $\therefore$ Inconsistent or dependent

There are a number of methods for finding the solutions of systems of linear equations. Before starting any method, check the slopes to determine whether or not to expect a solution.

The graphical method, which yields approximate solutions, consists simply of plotting the two straight lines and reading off the point of intersection.

Illustration. Solve
$$\begin{cases} x + y = 4 \\ 2y - 3x = 3 \end{cases}$$

Solution.   Obtain the intercepts

$$a_1 = 4, \quad b_1 = 4 \quad \text{and} \quad a_2 = -1, \quad b_2 = \frac{3}{2}$$

and draw the lines, Figure 23, p. 98. The point of intersection is $x = 1, y = 3$.

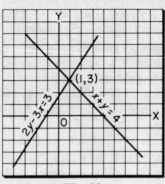

Fig. 23

Check. $1 + 3 = 4 = 4$

                    $2(3) - 3 = 3 = 3$

We shall now consider two algebraic methods, *elimination* and *substitution*.

Illustrations.

1) Solve

$$\begin{cases} 2x - 4y + 4 = 0 & (1) \\ 3x + 2y + 8 = 0 & (2) \end{cases}$$

Solution.

Write (1):	$2x - 4y + \ 4 = 0$	(3)
Multiply (2) by 2:	$6x + 4y + 16 = 0$	(4)
Add:	$8x \qquad + 20 = 0$	(5)
Solve (5) for x:	$x = -\tfrac{5}{2}$	(6)
Substitute $x = -\tfrac{5}{2}$ in (1):	$2(-\tfrac{5}{2}) - 4y + 4 = 0$	(7)
Solve (7) for y:	$y = -\tfrac{1}{4}$	

Solution is $(-\tfrac{5}{2}, -\tfrac{1}{4})$.

Check. Substitute into *both* (1) and (2).

$$2(-\tfrac{5}{2}) - 4(-\tfrac{1}{4}) + 4 = -5 + 1 + 4 = 0$$
$$3(-\tfrac{5}{2}) + 2(-\tfrac{1}{4}) + 8 = -\tfrac{15}{2} - \tfrac{1}{2} + 8 = -\tfrac{16}{2} + 8 = 0$$

2) Solve

$$\begin{cases} 2x - 4y + 4 = 0 & (1) \\ 3x + 2y + 8 = 0 & (2) \end{cases}$$

Solution.

Solve (1) for $x = f(y)$:	$x = 2y - 2$	(3)
Substitute (3) in (2):	$3(2y - 2) + 2y + 8 = 0$	(4)
Solve (4) for y:	$y = -\tfrac{1}{4}$	(5)
Substitute $y = -\tfrac{1}{4}$ in (1):	$2x - 4(-\tfrac{1}{4}) + 4 = 0$	(6)

Solve (6) for x: $x = -\frac{5}{2}$
Solution is $(-\frac{5}{2}, -\frac{1}{4})$.

In the first illustration we eliminated y by an addition after multiplying equation (2) through by a constant, 2; such multiplication does not change the equation. The system was thus reduced to one equation in one unknown, which could easily be solved. The second unknown was then found by a substitution. Frequently it is necessary to multiply both equations to effect an elimination.

Illustration.   Solve

$$\begin{cases} 3x - 2y = 6 & \text{(1)} \\ 2x - 3y = -1 & \text{(2)} \end{cases}$$

Solution.

Multiply (1) by 2:	$6x - 4y = 12$	(3)
Multiply (2) by 3:	$6x - 9y = -3$	(4)
Subtract (4) from (3):	$5y = 15$	(5)
Solve (5) for y:	$y = 3$	(6)
Multiply (1) by 3:	$9x - 6y = 18$	(7)
Multiply (2) by 2:	$4x - 6y = -2$	(8)
Subtract (8) from (7):	$5x = 20$	(9)
Solve (9) for x:	$x = 4$	

Solution is (4, 3).

Check.   Substitute into (1) and (2).

$$3(4) - 2(3) = 12 - 6 = 6$$
$$2(4) - 3(3) = 8 - 9 = -1$$

After we have obtained one value, the second can always be obtained by substituting into the original equations. However, if an error has been made in finding the first value, an erroneous value for the second variable will be obtained. It is therefore always necessary to check in both equations.

To solve a consistent system of three linear equations in three unknowns, we eliminate the same unknown from any two pairs of the equations and solve the resulting two linear equations in two unknowns. The value of the third unknown is then found by substituting into any of the original equations.

Illustration.   Solve

$$\begin{cases} 2x - 2y + z = 22 \\ 3x + y - 3z = 12 \\ x - 3y - z = 14 \end{cases}$$

Solution.   We shall eliminate y. Multiply the second equation by 2 and add to the first equation; multiply the second equation by 3 and add to the third equation.

$$8x - 5z = 46$$
$$\underline{10x - 10z = 50}$$

Eliminate z:
$$6x \qquad = 42$$
$$x = 7$$

Substitute:
$$5z = 8x - 46 = 56 - 46 = 10$$
$$z = 2$$

Substitute:
$$y = 12 + 3z - 3x = 12 + 6 - 21$$
$$y = -3$$

Check.
$$2(7) - 2(-3) + 2 = 14 + 6 + 2 = 22$$
$$3(7) + (-3) - 3(2) = 21 - 3 - 6 = 12$$
$$7 - 3(-3) - 2 = 7 + 9 - 2 = 14$$

The methods of elimination and substitution can be applied to a consistent system of linear equations in any number of unknowns.

57. Determinants. The symbol

$$\begin{vmatrix} a_1 & b_1 \\ a_2 & b_2 \end{vmatrix}$$

is called a **determinant of the second order.** It represents a number defined as follows:

$$\begin{vmatrix} a_1 & b_1 \\ a_2 & b_2 \end{vmatrix} = a_1 b_2 - a_2 b_1$$

Example:

$$\begin{vmatrix} 3 & 2 \\ 5 & -1 \end{vmatrix} = (3)(-1) - (2)(5) = -3 - 10 = -13$$

A determinant is a square array made up of
    elements, the quantities a_1, a_2, b_1, b_2;
    rows, the elements of a horizontal line;
    columns, the elements of a vertical line; and the
    principal diagonal, the diagonal starting in the upper left-hand
       corner.

To evaluate a *second-order* determinant, multiply the elements of the principal diagonal and from this product *subtract* the product of the elements of the other diagonal; schematically:

If a determinant has three rows and three columns, it is called a **third-order determinant.** The value of the determinant

$$\begin{vmatrix} a_1 & b_1 & c_1 \\ a_2 & b_2 & c_2 \\ a_3 & b_3 & c_3 \end{vmatrix}$$

is defined to be

$$a_1 b_2 c_3 + a_2 b_3 c_1 + a_3 b_1 c_2 - a_1 b_3 c_2 - a_3 b_2 c_1 - a_2 b_1 c_3$$

The following is a scheme for evaluating a third-order determinant: rewrite the first two columns to the right of the determinant; multiply the elements in each of the three diagonals running down from left to right and prefix the products with plus signs; then multiply the elements in each of the three diagonals running down from right to left and prefix these products with minus signs. The value of the determinant is the algebraic sum of these products. Symbolically,

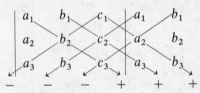

Example:

$$\begin{vmatrix} 3 & -1 & 5 \\ 1 & 2 & -2 \\ 2 & 3 & 6 \end{vmatrix} \begin{matrix} 3 & -1 \\ 1 & 2 \\ 2 & 3 \end{matrix} = (3)(2)(6) + (-1)(-2)(2) + (5)(1)(3)$$
$$- (-1)(1)(6) - (3)(-2)(3) - (5)(2)(2)$$
$$= 36 + 4 + 15 + 6 + 18 - 20 = 59$$

Determinants with n rows and n columns are called n^{th}-order determinants. The value of a determinant of order higher than 3 is found by expanding the determinant in terms of its minors. We shall not discuss determinants in this detail.*

Determinants may be used to solve systems of linear equations. Consider the general system

$$\begin{cases} a_1 x + b_1 y = c_1 \\ a_2 x + b_2 y = c_2 \end{cases}$$

Multiply the first equation by b_2, the second by $-b_1$, and add to

---

*See Kaj L. Nielsen, *College Mathematics,* New York: Barnes and Noble, Inc., 1958, p. 241; or R. A. Beaumont and R. W. Ball, *Introduction to Modern Algebra and Matrix Theory,* New York: Rinehart, 1954, Chapter I.

obtain

$$(a_1b_2 - a_2b_1)x = b_2c_1 - b_1c_2$$

The solution for x is

$$x = \frac{b_2c_1 - b_1c_2}{a_1b_2 - a_2b_1}$$

providing $a_1b_2 - a_2b_1 \neq 0$. If we eliminate x between the given equations and solve for y, we have

$$y = \frac{a_1c_2 - a_2c_1}{a_1b_2 - a_2b_1}$$

We notice that both denominators are the same. They can be represented by

$$D = \begin{vmatrix} a_1 & b_1 \\ a_2 & b_2 \end{vmatrix}$$

This determinant, D, is called the determinant of the system. For the equations to be consistent we must have $D \neq 0$.

The numerator of the solution for x is

$$b_2c_1 - b_1c_2 = \begin{vmatrix} c_1 & b_1 \\ c_2 & b_2 \end{vmatrix} = N_x$$

which is the determinant D with a_1 and a_2 (the coefficients of x) replaced by c_1 and c_2 (the constant terms). The numerator of the solution for y is

$$a_1c_2 - a_2c_1 = \begin{vmatrix} a_1 & c_1 \\ a_2 & c_2 \end{vmatrix} = N_y$$

which is D with b_1 and b_2 (the coefficients of y) replaced by c_1 and c_2. The solutions of the system are given by

$$x = \frac{N_x}{D} \quad \text{and} \quad y = \frac{N_y}{D}$$

Illustration. Solve the following system by determinants:

$$\begin{cases} 2x - 3y = 5 \\ x + 2y = 3 \end{cases}$$

Solution.

$$D = \begin{vmatrix} 2 & -3 \\ 1 & 2 \end{vmatrix} = 4 - (-3) = 7$$

$$N_x = \begin{vmatrix} 5 & -3 \\ 3 & 2 \end{vmatrix} = 10 - (-9) = 19$$

$$N_y = \begin{vmatrix} 2 & 5 \\ 1 & 3 \end{vmatrix} = 6 - 5 = 1$$

$$\therefore \qquad x = \frac{19}{7} \quad \text{and} \quad y = \frac{1}{7}$$

We have already noted that the equations are consistent only if $D \neq 0$. We have the additional properties:

a) If $D = 0$ and either $N_x \neq 0$ or $N_y \neq 0$, the equations are inconsistent.

b) If $D = 0$ and $N_x = N_y = 0$, the equations are dependent.

Consider the system

$$\begin{cases} a_1 x + b_1 y + c_1 z = d_1 \\ a_2 x + b_2 y + c_2 z = d_2 \\ a_3 x + b_3 y + c_3 z = d_3 \end{cases}$$

and let

$$D = \begin{vmatrix} a_1 & b_1 & c_1 \\ a_2 & b_2 & c_2 \\ a_3 & b_3 & c_3 \end{vmatrix} \neq 0$$

Then

$$x = \frac{1}{D} \begin{vmatrix} d_1 & b_1 & c_1 \\ d_2 & b_2 & c_2 \\ d_3 & b_3 & c_3 \end{vmatrix}$$

$$y = \frac{1}{D} \begin{vmatrix} a_1 & d_1 & c_1 \\ a_2 & d_2 & c_2 \\ a_3 & d_3 & c_3 \end{vmatrix}$$

$$z = \frac{1}{D} \begin{vmatrix} a_1 & b_1 & d_1 \\ a_2 & b_2 & d_2 \\ a_3 & b_3 & d_3 \end{vmatrix}$$

Illustration. Use determinants to solve

$$\begin{cases} x - 2y + 3z = 6 \\ 3x - y - z = -2 \\ 2x - 3y + 2z = 2 \end{cases}$$

Solution.

$$D = \begin{vmatrix} 1 & -2 & 3 \\ 3 & -1 & -1 \\ 2 & -3 & 2 \end{vmatrix} = -10$$

$$N_x = \begin{vmatrix} 6 & -2 & 3 \\ -2 & -1 & -1 \\ 2 & -3 & 2 \end{vmatrix} = -10$$

$$N_y = \begin{vmatrix} 1 & 6 & 3 \\ 3 & -2 & -1 \\ 2 & 2 & 2 \end{vmatrix} = -20$$

$$N_z = \begin{vmatrix} 1 & -2 & 6 \\ 3 & -1 & -2 \\ 2 & -3 & 2 \end{vmatrix} = -30$$

$$\therefore \quad x = \frac{-10}{-10} = 1, \quad y = \frac{-20}{-10} = 2, \quad z = \frac{-30}{-10} = 3$$

58. Matrices. A rectangular array of numbers written in the form

$$A = \begin{pmatrix} a_{11} & a_{12} & a_{13} & \ldots & a_{1n} \\ a_{21} & a_{22} & a_{23} & \ldots & a_{2n} \\ . & . & . & \ldots & . \\ a_{m1} & a_{m2} & a_{m3} & \ldots & a_{mn} \end{pmatrix}$$

is called a **matrix.** This matrix has m rows and n columns and is referred to as an $m \times n$ matrix. If $m = n$, the matrix is **square.** The elements a_{ij} represent numbers.

Examples:

  a) a 1×3 matrix: $(3, -5, 7)$

  b) a 3×1 matrix: c) a 2×3 matrix:

$$\begin{pmatrix} 7 \\ 3 \\ 2 \end{pmatrix} \qquad\qquad \begin{pmatrix} 6 & 0 & 2 \\ 1 & -5 & 7 \end{pmatrix}$$

  d) a 4×4 matrix: e) a 4×2 matrix:

$$\begin{pmatrix} 3 & 6 & -1 & 8 \\ 1 & 7 & -3 & 9 \\ 2 & 8 & -5 & 4 \\ 4 & 9 & -7 & 2 \end{pmatrix} \qquad \begin{pmatrix} a & e \\ b & f \\ c & g \\ d & h \end{pmatrix}$$

We emphasize that a matrix does *not* represent a single number, as a determinant does.

Definition. *Two matrices are* **equal** *if and only if they have the same dimensions and each element of one is identically equal to the corresponding element of the other.* Thus

$$\begin{pmatrix} a & b \\ c & d \end{pmatrix} = \begin{pmatrix} w & x \\ y & z \end{pmatrix}$$

if and only if

$$a = w, b = x, c = y, d = z$$

Definition. *The* **sum** *of two matrices of the same dimensions is a matrix whose elements are the sum of the corresponding elements of the two given matrices.*

$$\begin{pmatrix} a & b \\ c & d \end{pmatrix} + \begin{pmatrix} w & x \\ y & z \end{pmatrix} = \begin{pmatrix} a + w & b + x \\ c + y & d + z \end{pmatrix}$$

Definition. *The* **product** *of a number* k *times a matrix is a matrix whose elements are* k *times the corresponding elements of the matrix.*

$$k\begin{pmatrix} a & b \\ c & d \end{pmatrix} = \begin{pmatrix} ka & kb \\ kc & kd \end{pmatrix}$$

Definition. *If* A *is an* m × n *matrix and* B *is an* n × p *matrix, then the* **product** AB = C *is an* m × p *matrix whose elements are the sum of the products of the elements of each row of* A *by the elements of each column of* B *in the proper order.*

$$\begin{pmatrix} a_1 & a_2 & a_3 \\ b_1 & b_2 & b_3 \end{pmatrix} \begin{pmatrix} x_1 & x_2 \\ y_1 & y_2 \\ z_1 & z_2 \end{pmatrix}$$

$$= \begin{pmatrix} a_1x_1 + a_2y_1 + a_3z_1 & a_1x_2 + a_2y_2 + a_3z_2 \\ b_1x_1 + b_2y_1 + b_3z_1 & b_1x_2 + b_2y_2 + b_3z_2 \end{pmatrix}$$

Examples:

a) $\begin{pmatrix} 3 & -1 \\ 5 & 2 \end{pmatrix} + \begin{pmatrix} -1 & 2 \\ -3 & 1 \end{pmatrix} = \begin{pmatrix} 2 & 1 \\ 2 & 3 \end{pmatrix}$

b) $\begin{pmatrix} 6 & 2 \\ 7 & 3 \\ 1 & 1 \end{pmatrix} - 3\begin{pmatrix} 2 & 1 \\ 1 & 3 \\ 0 & -2 \end{pmatrix} = \begin{pmatrix} 0 & -1 \\ 4 & -6 \\ 1 & 7 \end{pmatrix}$

c) $\begin{pmatrix} 4 & -1 \\ 2 & 1 \end{pmatrix} \begin{pmatrix} 2 & -3 \\ 1 & -2 \end{pmatrix} = \begin{pmatrix} 7 & -10 \\ 5 & -8 \end{pmatrix}$

d) $\begin{pmatrix} 5 & -1 \\ 2 & 2 \\ 1 & 3 \end{pmatrix} \begin{pmatrix} 1 & -2 & 1 \\ 3 & 3 & -2 \end{pmatrix} = \begin{pmatrix} 2 & -13 & 7 \\ 8 & 2 & -2 \\ 10 & 7 & -5 \end{pmatrix}$

Note that the number of columns of A must equal the number of rows of B for the product AB to exist. Both products AB and BA exist only if A and B are square matrices. However, $AB \neq BA$; i.e., multiplication of square matrices is not commutative. Multiplication of matrices is associative, $(AB)C = A(BC)$.

Example:

$$\begin{pmatrix} 3 & -1 \\ 1 & 2 \end{pmatrix} \begin{pmatrix} 1 & 2 \\ 3 & 1 \end{pmatrix} \begin{pmatrix} 4 & 1 & 3 \\ 2 & 3 & 1 \end{pmatrix} = \begin{pmatrix} 0 & 5 \\ 7 & 4 \end{pmatrix} \begin{pmatrix} 4 & 1 & 3 \\ 2 & 3 & 1 \end{pmatrix}$$

$$= \begin{pmatrix} 10 & 15 & 5 \\ 36 & 19 & 25 \end{pmatrix}$$

$$= \begin{pmatrix} 3 & -1 \\ 1 & 2 \end{pmatrix} \begin{pmatrix} 8 & 7 & 5 \\ 14 & 6 & 10 \end{pmatrix}$$

Definition. *A square matrix whose elements are all* 0 *except those of the main diagonal which are all unity is called the multiplicative* **identity matrix** *and is denoted by* **I**.

It is easily seen that $AI = IA = A$:

$$\begin{pmatrix} a_1 & b_1 \\ a_2 & b_2 \end{pmatrix} \begin{pmatrix} 1 & 0 \\ 0 & 1 \end{pmatrix} = \begin{pmatrix} a_1 & b_1 \\ a_2 & b_2 \end{pmatrix}$$

Definition. *The determinant whose elements are identically those of a square matrix is called the* **determinant of the matrix.**

If the determinant of a square matrix A is not equal to 0, there exists a matrix A^{-1} such that $A^{-1}A = AA^{-1} = I$. The matrix A^{-1} is called the **inverse** of A. We shall calculate the inverse for a 2×2 matrix. Let

$$A = \begin{pmatrix} a & b \\ c & d \end{pmatrix} \quad \text{and} \quad A^{-1} = \begin{pmatrix} w & x \\ y & z \end{pmatrix}$$

We can find the elements of A^{-1} in terms of those of A by solving the matrix equation

$$\begin{pmatrix} a & b \\ c & d \end{pmatrix} \begin{pmatrix} w & x \\ y & z \end{pmatrix} = \begin{pmatrix} 1 & 0 \\ 0 & 1 \end{pmatrix}$$

Perform the multiplication on the left to obtain

$$\begin{pmatrix} aw + by & ax + bz \\ cw + dy & cx + dz \end{pmatrix} = \begin{pmatrix} 1 & 0 \\ 0 & 1 \end{pmatrix}$$

These two matrices are equal if and only if

$$aw + by = 1 \qquad ax + bz = 0$$
$$cw + dy = 0 \qquad cx + dz = 1$$

Let $\Delta = ad - bc \neq 0$ and solve the two systems to obtain

$$w = \frac{d}{\Delta}, \; x = -\frac{b}{\Delta}, \; y = -\frac{c}{\Delta}, \; z = \frac{a}{\Delta}$$

and finally

$$A^{-1} = \frac{1}{\Delta}\begin{pmatrix} d & -b \\ -c & a \end{pmatrix}$$

Illustration. Find the inverse of $A = \begin{pmatrix} 3 & -1 \\ 2 & 5 \end{pmatrix}$

Solution. The determinant of A is

$$\Delta = \begin{vmatrix} 3 & -1 \\ 2 & 5 \end{vmatrix} = (3)(5) - (2)(-1) = 17$$

The inverse is then given by

$$A^{-1} = \frac{1}{17}\begin{pmatrix} 5 & 1 \\ -2 & 3 \end{pmatrix}$$

Systems of equations may be conveniently written in matrix notation. Consider the system

$$\begin{cases} a_1 x + b_1 y + c_1 = 0 \\ a_2 x + b_2 y + c_2 = 0 \end{cases}$$

and let

$$A = \begin{pmatrix} a_1 & b_1 \\ a_2 & b_2 \end{pmatrix}, \quad X = \begin{pmatrix} x \\ y \end{pmatrix}, \quad C = \begin{pmatrix} c_1 \\ c_2 \end{pmatrix}, \quad 0 = \begin{pmatrix} 0 \\ 0 \end{pmatrix}$$

The system may then be written

$$AX + C = 0$$

If we now treat this as a simple linear equation, we can perform the following operations:

$$AX = -C$$
$$A^{-1}AX = -A^{-1}C$$
$$IX = -A^{-1}C \qquad [\text{Since } A^{-1}A = I]$$
$$X = -A^{-1}C \qquad [\text{Since } IX = X]$$

Thus the solution can be obtained in terms of C and the inverse of A.

Illustration.   Solve the system

$$\begin{cases} 3x - 2y + 1 = 0 \\ x + 5y - 3 = 0 \end{cases}$$

Solution.   In terms of matrices we have

$$A = \begin{pmatrix} 3 & -2 \\ 1 & 5 \end{pmatrix}, \quad \Delta = \begin{vmatrix} 3 & -2 \\ 1 & 5 \end{vmatrix} = 17$$

$$A^{-1} = \frac{1}{17} \begin{pmatrix} 5 & 2 \\ -1 & 3 \end{pmatrix}, \quad C = \begin{pmatrix} 1 \\ -3 \end{pmatrix}$$

and

$$X = -A^{-1}C = -\frac{1}{17} \begin{pmatrix} 5 & 2 \\ -1 & 3 \end{pmatrix} \begin{pmatrix} 1 \\ -3 \end{pmatrix} = -\frac{1}{17} \begin{pmatrix} -1 \\ -10 \end{pmatrix} = \begin{pmatrix} x \\ y \end{pmatrix}$$

Therefore $x = \dfrac{1}{17}$ and $y = \dfrac{10}{17}$

Although matrices provide an elegant method for solving systems of linear equations, the calculation of the inverse is quite cumbersome for systems with more than two unknowns. In practical problems this calculation is performed by electronic calculators.

59.   Linear Inequalities.   The expressions

$$\boldsymbol{ax + b > 0} \qquad \text{and} \qquad \boldsymbol{ax + b < 0}$$

represent **linear inequalities in one variable**. The solution of an inequality is that set of numbers which satisfies the inequality. To find this set of numbers for any given inequality, we use the properties of operations given in Section 33.

Illustration.   Solve $4x + 3 < x - 9$.
Solution.   Subtract $x + 3$ from both sides of the inequality to obtain $3x < -12$; now divide both sides by 3 to obtain $x < -4$. The solution set is the set of numbers whose elements are less than -4. Another way to write this is in terms of the open interval $]-\infty, -4[$.

The equality sign may be used in conjunction with the inequality sign, $\leq$, read, "less than or equal to."

Illustration.   Solve $4x + 3 \leq x - 9$.
Solution.   This is the same problem as the above illustration except that the equal sign is included. We perform the same steps to obtain $x \leq -4$. The solution now includes the end point -4 and the interval is $]-\infty, -4]$.

The expressions

$$ax + by + c < 0 \quad \text{and} \quad ax + by + c > 0$$

represent **linear inequalities in two variables**. The solution is the set of ordered pairs (x, y) which satisfies the given inequality. This set can be discussed geometrically. Consider first the fact that the linear equation $ax + by + c = 0$ is represented by a straight line which divides the plane into two half-planes. If $a \neq 0$, we have a *right half-plane* and a *left half-plane*. Consider a point P with coordinates (x_0, y_0) and draw the line PQ parallel to the X-axis and intersecting $ax + by + c = 0$ at the point (x_1, y_0).

Definition. *The point* $P(x_0, y_0)$ *lies in the* **right half-plane** *if* $x_0 > x_1$; *it lies in the* **left half-plane** *if* $x_0 < x_1$.

If $a = 0$ the line $ax + by + c = 0$ is parallel to the X-axis. Draw a line PQ from $P(x_0, y_0)$ parallel to the Y-axis intersecting this line at $Q(x_0, y_1)$.

Definition. *The point* $P(x_0, y_0)$ *lies in the* **upper half-plane** *if* $y_0 > y_1$ *and in the* **lower half-plane** *if* $y_0 < y_1$.

We can now describe the graph of a set of pairs which satisfies an inequality $\{(x, y) | ax + by + c > 0\}$. The procedure is described in the following illustration.

Illustration. Describe the graph of

$$3x - y - 6 > 0$$

Solution. The intercepts of the line

$$3x - y - 6 = 0$$

are $(0, -6)$ and $(2, 0)$; draw the line through these points, Figure 24. Choose any point (the easiest is the origin) and test it in the inequality. The point $(0, 0)$ yields $-6 > 0$, which is not true, so the origin does *not* satisfy the inequality. Since the origin is to the *left* of the line, all the points to the left of the line do *not* satisfy the inequality. Consequently, the graph of the inequality is the set of points in the right half-plane which has been shaded in Figure 24, p. 110. As a check, test a point in the right half-plane. [The point $(3, 0)$ yields $9 - 6 = 3 > 0$.]

A system of simultaneous linear inequalities in two variables has for its solution that set of ordered pairs which satisfies all of the inequalities in the system. Consider the system

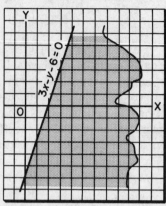

Fig. 24

$$\begin{cases} a_1x + b_1y + c_1 > 0 \\ a_2x + b_2y + c_2 > 0 \end{cases}$$

This system is satisfied by the set

$$\{(x, y)|a_1x + b_1y + c_1 > 0\} \cap \{(x, y)|a_2x + b_2y + c_2 > 0\}$$

Graphically, we look for those points which lie in both of the two given half-planes.

Illustrations.

1) Solve the system

$$\begin{cases} 2x - y + 6 > 0 \\ x - 2y + 4 > 0 \end{cases}$$

Solution. Draw the two lines obtained by replacing the inequality signs by equal signs. The intersection of the two half-planes is the set of points shown by the shaded area of Figure 25.

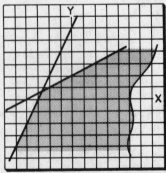

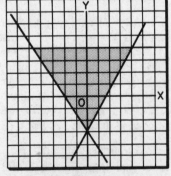

Fig. 25   Fig. 26

2) Solve the system

$$\begin{cases} y - 4 < 0 \\ 3x + 2y + 6 > 0 \\ 7x - 4y - 12 < 0 \end{cases}$$

Solution. The solution set is shown as the shaded area of Figure 26.

60. Linear Programming.

In recent years the linear function has found great application to business and managerial problems. The technique has been given the name *linear programming.* Although the general problem is very complex and requires the use of large-scale computers, we can explain the idea by considering the simple problem with two variables.*

The Two-Variable Problem. Find the maximum and/or minimum value of a linear function $f: ax + by + c$, constrained by a system of inequalities in x and y.

We saw in the last section that the solution of a linear inequality is the set of points in the half-plane which satisfies the inequality. If the equality is also stated, we include the boundary (the line itself) and call it a *closed* half-plane. The set of points which satisfies a system of inequalities (with the equality included) is the intersection of the closed half-planes; this intersection is called a **polygonal convex set.** If this set has a finite area, we call it a **convex polygon.** We now state the following theorem.

Theorem. *The function* $f: ax + by + c$ *defined over a convex polygon takes on its maximum (minimum) value at a vertex point of the convex polygon.*

The solution to the simple two-variable linear programming problem is obtained by finding the vertices of the convex polygon, substituting the coordinates of these vertices into the function, and establishing the largest and smallest values.

Illustration. Find the extrema of the linear function $f: 2x - 3y + 5$ defined over the convex polygon which is the solution set of the system

$$\begin{cases} y - 4 \le 0 \\ 3x + 2y + 6 \ge 0 \\ 7x - 4y - 12 \le 0 \end{cases}$$

---

*For a more detailed discussion and additional bibliography see Kaj L. Nielsen, *Methods in Numerical Analysis,* 2nd ed., New York: Macmillan, 1964, Chapter 9.

Solution. Draw the three lines defined by the equations in the system and find the vertices of the polygon. See Figure 26. The vertices are $(0, -3)$, $(4, 4)$, $(-14/3, 4)$. Substitute these values into the function f to obtain.

$$f(0, -3) = 0 + 9 + 5 = 14 \quad \text{(maximum)}$$
$$f(4, 4) = 8 - 12 + 5 = 1$$
$$f\left(-\frac{14}{3}, 4\right) = -\frac{28}{3} - 12 + 5 = -\frac{49}{3} \quad \text{(minimum)}$$

Summary of Symbols

$mx + b$, linear expression in one variable

$ax + by + c$, linear expression in two variables

$a_1x_1 + a_2x_2 + \ldots + a_nx_n$, linear expression in n variables

Δx, increment of x

$$\begin{vmatrix} a_1 & b_1 & c_1 \\ a_2 & b_2 & c_2 \\ a_3 & b_3 & c_3 \end{vmatrix}, \text{ third-order determinant}$$

$$\begin{pmatrix} a_1 & b_1 & c_1 & d_1 \\ a_2 & b_2 & c_2 & d_2 \\ a_3 & b_3 & c_3 & d_3 \end{pmatrix}, 3 \times 4 \text{ matrix}$$

$$I = \begin{pmatrix} 1 & 0 & 0 \\ 0 & 1 & 0 \\ 0 & 0 & 1 \end{pmatrix}, \text{ identity matrix}$$

A^{-1}, inverse matrix of A

$AX + C = 0$, system of equations in matrix form.

---

61. Exercise VI.

1. Draw the graph and find the range of each of the linear functions defined by the following equations:

   a) $y = 3x - 2$, whose domain is $x \in R$

   b) $2x - 4y + 3 = 0$, whose domain is $0 \leq x \leq 6$

   c) $3y - x = 18$, whose domain is $[0, 6]$

   d) $y = mx + b$, if $m = -1$, $b = 3$, and $-3 \leq x \leq 0$

   e) $v = 32t + 10$, if $0 \leq t \leq 10$

2. Find the slopes and intercepts of the straight lines defined by the following linear equations:

   a) $2y - 4x = 5$ b) $3x + 6y - 1 = 0$

   c) $x + y = 7$ d) $d = 2t + 8$

   e) $6w + l = 24$ f) $5x - 10y = 7$

3. Find the slope and equation of the line through each of the following pairs of points:
   a) $P_1(-3, 4)$ and $P_2(2, -1)$
   b) $P_1(0, 3)$ and $P_2(3, 0)$
   c) $P_1(-1, -4)$ and $P_2(-2, -5)$
   d) $P_1(0, 0)$ and $P_2(3, -1)$
   e) $P_1(3, 3)$ and $P_2(-1, -1)$
   f) $P_1(-2, 3)$ and $P_2(-1, 0)$
   g) $P_1(2, 1)$ and $P_2(5, 1)$
   h) $P_1(3, 4)$ and $P_2(3, 2)$

4. Solve the following systems of linear equations, first by elimination and substitution and then by determinants:

   a) $\begin{cases} 3x - 5y = 4 \\ x + y = 2 \end{cases}$

   b) $\begin{cases} 2x - 3y - 3 = 0 \\ 2y - 3x + 5 = 0 \end{cases}$

   c) $\begin{cases} ax + 26 = -4 \\ 2x - by = 3 \end{cases}$

   d) $\begin{cases} 2y - 5x = 6 \\ 3x + 2y = 2 \end{cases}$

   e) $\begin{cases} x + y + z = 1 \\ 3x - 2y - z = 5 \\ x + 3y + 2z = 3 \end{cases}$

   f) $\begin{cases} x - y + z - 1 = 0 \\ 6x + 5y - z + 3 = 0 \\ x - 2y + 2z - 5 = 0 \end{cases}$

   g) $\begin{cases} 32x - 17y + 15z = -2 \\ -3x + y + 5z = -8 \\ 16x + 7y - 10z = 63 \end{cases}$

   h) $\begin{cases} x + 2z = 1 \\ x - 3y = 2 \\ y + z = 3 \end{cases}$

5. Find the resulting matrices:

   a) $\begin{pmatrix} 3 & -6 \\ 4 & 0 \end{pmatrix} + \begin{pmatrix} 1 & 1 \\ 1 & 1 \end{pmatrix}$

   b) $\begin{pmatrix} 2 & 1 \\ 3 & -2 \\ 1 & -5 \end{pmatrix} - \begin{pmatrix} 6 & -1 \\ 2 & -3 \\ 4 & 2 \end{pmatrix}$

   c) $(3, 1, 5) + \begin{pmatrix} 1 \\ 3 \\ 2 \end{pmatrix}$

   d) $\begin{pmatrix} 2 & 1 & -1 \\ 5 & 2 & 4 \end{pmatrix} - 3 \begin{pmatrix} 1 & 0 & -1 \\ 2 & 3 & 1 \end{pmatrix}$

   e) $2 \begin{pmatrix} 3 & -1 \\ 2 & 6 \end{pmatrix} + 4 \begin{pmatrix} -1 & 0 \\ 2 & -3 \end{pmatrix}$

f) $2\begin{pmatrix} a_1 & b_1 \\ c_1 & d_1 \end{pmatrix} + \begin{pmatrix} a_2 & b_2 \\ c_2 & d_1 \end{pmatrix}$

g) $\begin{pmatrix} 6 & -1 \\ 1 & 3 \end{pmatrix} + \begin{pmatrix} 1 & 0 \\ 0 & 1 \end{pmatrix}$

h) $\begin{pmatrix} 3 & -1 \\ 1 & 2 \\ 2 & 6 \end{pmatrix} \begin{pmatrix} 2 & 1 \\ 1 & 2 \end{pmatrix}$

i) $(1, -1, 1)\begin{pmatrix} 3 \\ 2 \\ 1 \end{pmatrix}$

j) $3\begin{pmatrix} 2 & -1 & 3 \\ 1 & 0 & 1 \\ 4 & 3 & 0 \end{pmatrix} \begin{pmatrix} 1 & 4 & 0 \\ 0 & -1 & 1 \\ 2 & -3 & 2 \end{pmatrix}$

k) $\begin{pmatrix} 3 & 0 & -1 & 2 \\ 1 & -3 & 1 & -1 \end{pmatrix} \begin{pmatrix} 1 & 3 & 2 \\ 2 & -1 & 3 \\ 0 & 0 & 1 \\ 1 & -1 & 0 \end{pmatrix}$

l) $\begin{pmatrix} 3 & 1 & 0 \\ 1 & -1 & 2 \end{pmatrix} \begin{pmatrix} 6 \\ -1 \\ 2 \end{pmatrix} (3 \quad 0 \quad 1 \quad -2)$

6. Solve the system $AX + C = 0$ by using the inverse matrix A^{-1}:

a) $A = \begin{pmatrix} 3 & -1 \\ 2 & 1 \end{pmatrix}, C = \begin{pmatrix} 6 \\ 9 \end{pmatrix}$

b) $A = \begin{pmatrix} 2 & -1 \\ 1 & 7 \end{pmatrix}, C = \begin{pmatrix} 0 \\ 1 \end{pmatrix}$

c) $A = \begin{pmatrix} 7 & 9 \\ 6 & 11 \end{pmatrix}, C = \begin{pmatrix} 5 \\ 7 \end{pmatrix}$

7. Solve the following inequalities:
   a) $3x + 7 > 0$ b) $5 - x < 0$
   c) $2x \geq 16$ d) $3x - 16 \leq 2$

8. Find graphically the solution sets of the following systems of inequalities:

a) $\begin{cases} x - 3y + 6 \geq 0 \\ y \geq 0 \\ x + 3y - 6 \leq 0 \end{cases}$

b) $\begin{cases} y + x - 5 \geq 0 \\ 3y - 4x - 8 \geq 0 \\ x + y - 12 \leq 0 \\ 3y - 4x + 6 \leq 0 \end{cases}$

c) $\begin{cases} x \le 3 \\ y \le 3 \\ y \ge -2 \\ 3y - 5x \ge 14 \\ 3y + 5x \le 24 \end{cases}$ d) $\begin{cases} x \ge -4 \\ x \le 4 \\ y \ge -4 \\ y \le 4 \\ y \ge x + 4 \end{cases}$

9. Find the maximum and minimum of each of the following linear functions, f, defined over each of the sets of Problem 8 :
   a) $8x - 12y + 3$ b) $1.20x + 2.50y - 6.75$

10. The difference of three times a number and a second number is 2. Find the numbers if their sum is 10.

VI

THE ALGEBRA OF POLYNOMIALS

62. Introduction. The polynomial is one of the most useful algebraic expressions in both pure and applied mathematics. We introduced the polynomial equation in one variable in Section 20 and discussed the linear equation in the last chapter. We now turn our attention to polynomials of higher degree than one. We shall discuss the equations, inequalities, and functions defined by polynomial equations, with their graphical representations. At the same time we shall introduce some useful algebraic tools.

63. The Quadratic Equation. A quadratic equation is one in which the highest degree of the variable is two. The equation

$$ax^2 + bx + c = 0, a \neq 0$$

where a, b, and c are constants, is called the **general quadratic equation in** x. If $a = 0$ the equation reduces to a linear equation; if $b = 0$ the equation is called a *pure* quadratic equation. We shall consider two methods for finding the solutions of quadratic equations.

I. *Solution by factoring.* If a product equals 0, one or more of its factors equals 0. If $XY = 0$, then $X = 0$ and/or $Y = 0$. To solve the quadratic equation by factoring, first factor the quadratic into its linear factors, then set each factor equal to 0 and solve the resulting linear equations.

Illustration. Solve $2x^2 - 3x = 9$.

Solution. Transpose: $2x^2 - 3x - 9 = 0$

Factor: $(2x + 3)(x - 3) = 0$

Equate each factor with 0:

$$2x + 3 = 0 \quad \Big| \quad x - 3 = 0$$

Solve for x: $\quad x = -\dfrac{3}{2} \quad \Big| \quad x = 3$

Check. $\quad 2\left(\dfrac{9}{4}\right) - 3\left(-\dfrac{3}{2}\right) = \dfrac{9}{2} + \dfrac{9}{2} = 9$

116

$$2(9) - 3(3) = 18 - 9 = 9$$

II. *Solution by the quadratic formula.* We shall derive the quadratic formula by a process known as **completing the square**. This process is itself a useful algebraic tool and deserves careful study. Consider the general quadratic equation.

1) Transpose the constant term to the right side:

$$ax^2 + bx = -c$$

2) Divide by the coefficient of x^2:

$$x^2 + \frac{b}{a}x = -\frac{c}{a}$$

3) Take 1/2 the coefficient of x, square it, and add to both sides:

$$x^2 + \frac{b}{a}x + \left(\frac{b}{2a}\right)^2 = \frac{b^2}{4a^2} - \frac{c}{a}$$

4) Factor the left side and simplify the right side:

$$\left(x + \frac{b}{2a}\right)^2 = \frac{b^2 - 4ac}{4a^2}$$

5) Extract the square root of each member:

$$x + \frac{b}{2a} = \frac{\pm \sqrt{b^2 - 4ac}}{2a}$$

6) Solve for x:

$$x = \frac{-b \pm \sqrt{b^2 - 4ac}}{2a}$$

This is known as the **quadratic formula**. At Step 4 we made the left side a perfect square and finished the process of completing the square. The quadratic formula may be used to solve a quadratic equation by substituting the given coefficients into the formula and simplifying.

Illustration.   Solve $2x^2 + 3x - 1 = 0$.
Solution. $a = 2$, $b = 3$, $c = -1$

$$x = \frac{-b \pm \sqrt{b^2 - 4ac}}{2a}$$

$$= \frac{-3 \pm \sqrt{9 - 4(2)(-1)}}{2(2)}$$

$$= \frac{-3 \pm \sqrt{9 + 8}}{4}$$

$$= \frac{-3 \pm \sqrt{17}}{4}$$

The expression $b^2 - 4ac$ which occurs under the radical in the quadratic formula is called the **discriminant** of the quadratic equation. It gives some useful information about the roots which we shall summarize in the following tables.

If the coefficients a, b, *and* c *are real numbers and*

if	then the roots of $ax^2 + bx + c = 0$ are
$b^2 - 4ac > 0$	real and unequal
$b^2 - 4ac = 0$	real and equal
$b^2 - 4ac < 0$	not real

If the roots are real and the coefficients are rational numbers, and

if	the roots are
$b^2 - 4ac$ is a perfect square	rational
$b^2 - 4ac$ is not a perfect square	irrational

Let

$$r_1 = \frac{-b + \sqrt{b^2 - 4ac}}{2a} \quad \text{and} \quad r_2 = \frac{-b - \sqrt{b^2 - 4ac}}{2a}$$

Then it is easily shown that

$$r_1 + r_2 = -\frac{b}{a} \quad \text{and} \quad r_1 \cdot r_2 = \frac{c}{a}$$

Some special cases are discussed in the following illustrations.

Illustrations.
1) Solve $3x^2 - 6x = 0$.
Solution. Factor to obtain

$$3x (x - 2) = 0$$

Then $3x = 0$ and $x - 2 = 0$

or $x = 0$ and $x = 2$

2) Solve $x^2 - 2x + 1 = 0$.

Solution. Factor to obtain

$$(x - 1)^2 = 0$$

Then $x = 1$ (twice)

3) Solve $4x^2 = 9$.

Solution. Solve for x^2 to obtain $x^2 = \dfrac{9}{4}$

Take the square root of both sides. $x = \pm \dfrac{3}{2}$

64. Quadratics in Two Variables. The general quadratic equation in two variables may be written in the form

$$ax^2 + bxy + cy^2 + dx + ey + f = 0$$

where a, b, c, d, e, and f are constants and a, b, and c are not all 0, since then the equation would reduce to a linear equation. There is an infinite number of pairs (x, y) which satisfy this equation. The graph of the function defined by such an equation with the elements in the real-number system is a curve. Two of these equations considered simultaneously form a system of quadratic equations which has four solutions. Although the solution set of a system of quadratic equations can be found algebraically, the work may be very cumbersome, since it involves the algebraic solution of a fourth-degree equation.* However, there are six cases that can be handled by special techniques. These cases are

a) one equation is linear;
b) $bxy + f = 0$ and $ax^2 + cy^2 + f = 0$;
c) both equations are of the form $ax^2 + cy^2 + f = 0$;
d) $ax^2 + cy^2 + f = 0$ and $ax^2 + cy + f = 0$ or
   $cy^2 + dx + f = 0$;
e) both equations are of the form $ax^2 + bxy + cy^2 + f = 0$;
f) both equations are of the form

$$A(x^2 + y^2) + Bxy + D(x + y) + F = 0$$

The procedure for each of these cases will be illustrated in Exercise VII.

---

*The solution set of a system of nonlinear equations is usually obtained by an iteration. See Kaj L. Nielsen, *Methods in Numerical Analysis,* 2nd ed., New York: Macmillan, 1964, Chapter 6.

If the solution set of a quadratic system is a subset of the real-number system, its members may be approximated graphically. We shall now turn our attention to the interesting subject of the graphs of quadratic equations in two variables.

65. The Graphs of Quadratics. The graph of the quadratic function

$$\{(x, y) | y = ax^2 + bx + c, a \neq 0\}$$

is a parabola, a smooth curve which rises or falls to a peak. If $a > 0$, the curve opens *upward* and if $a < 0$, the curve opens *downward*. The two cases are shown in Figures 27 and 28.

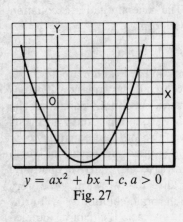

$y = ax^2 + bx + c, a > 0$
Fig. 27

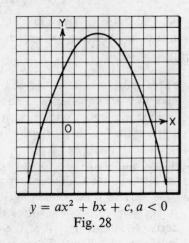

$y = ax^2 + bx + c, a < 0$
Fig. 28

The graph of the relation defined by

$$\{(x, y) | y^2 = 4ax, a \neq 0\}$$

is also a parabola. (Let $a = 1$. Calculate a table of ordered pairs and draw the graph.)

Both of the above examples are special cases of the general quadratic relation

$$\{(x, y) | Ax^2 + Bxy + Cy^2 + Dx + Ey + F = 0\}$$

The graph of this relation is one of a group of curves called the *conics*. The parabola is one such curve. Although there is more than one way to define a conic, we shall use the following definition.*

---

*The conics can also be defined geometrically. See Kaj L. Nielsen, *College Mathematics*, New York: Barnes and Noble, Inc., 1958, Chapter 7.

Definition. *A* **conic** *is a curve whose equation is of the second degree.*

A complete classification of the conics can be made in terms of the quadratic relation.

If the given quadratic relations can be transformed into the form	the conic is called
$(y - k)^2 = 4a(x - h)$ $(x - h)^2 = 4a(y - k)$	a **parabola**
$\dfrac{(x - h)^2}{a^2} + \dfrac{(y - k)^2}{b^2} = 1$	an **ellipse**
$(x - h)^2 + (y - k)^2 = r^2$	a **circle**
$\dfrac{(x - h)^2}{a^2} - \dfrac{(y - k)^2}{b^2} = \pm 1$ $xy = K$	a **hyperbola**

In the case of the parabola the point (h, k) is called the **vertex**; in the other cases it is called the **center**. It should be noted that the circle is a special case of the ellipse $(a = b = r)$. If the point (h, k) is the origin $(h = k = 0)$, the equations are very simple. Although it is possible to make $(h, k) = (0, 0)$ by a translation and rotation of axes, we can determine the nature of the graph by considering the general equation directly. Consider the general equation of the second degree in two variables

$$Ax^2 + Bxy + Cy^2 + Dx + Ey + F = 0$$

and define the discriminant of this equation to be the determinant

$$\Delta = \begin{vmatrix} 2A & B & D \\ B & 2C & E \\ D & E & 2F \end{vmatrix}$$

If $\Delta = 0$, the conic is *degenerate* and the graph may be parallel lines, intersecting lines, or a point. If $\Delta \neq 0$, the graph is a proper conic and

if	the graph is
$B^2 - 4AC = 0$	a parabola
$B^2 - 4AC < 0$	an ellipse
$B^2 - 4AC > 0$	a hyperbola

Examples:

a) $4x^2 - x + y - 3 = 0$

$A = 4, B = C = 0, D = -1, E = 1, F = -3$

$B^2 - 4AC = 0 - 0 = 0$ The graph is a parabola.

b) $5x^2 + 2xy + 3y^2 - 7 = 0$

$A = 5, B = 2, C = 3, D = E = 0, F = -7$

$B^2 - 4AC = 4 - 60 = -56 < 0$

The graph is an ellipse.

c) $2x^2 - 3xy - y^2 + 3x = 0$

$A = 2, B = -3, C = -1, D = 3, E = F = 0$

$B^2 - 4AC = 9 + 8 = 17 > 0$

The graph is a hyperbola.

Although the graphs can always be obtained by calculating a table of values, the calculations may be too time-consuming. A good sketch can be obtained by noting some characteristics of the conics.

In the following we shall let $B = 0$. The xy term can be removed by a rotation of the axes, a technique that requires some knowledge of trigonometry.

The Parabola. There are two characteristic points for the parabola, the vertex, $V(h, k)$, and the focus, F. The line through V and F is the axis of the parabola and the curve is symmetric* with respect to this axis. Let the line through F and perpendicular to the axis intersect the parabola at points A and B.

Illustrations.

1) Find the characteristic points of the parabola

$$32y = 4x^2 - 12x + 89$$

Solution. Divide the equation by 4 and transpose:

Complete the square: $\left(x - \dfrac{3}{2}\right)^2 = 8y - \dfrac{89}{4} + \dfrac{9}{4}$

---

*If a line is the perpendicular bisector of a segment between two points A and B, it is called an *axis of symmetry*, and the points are said to be *symmetric* with respect to the line.

Equation	F	A	B	Curve Opening
Axis Parallel to X-axis				
$(y - k)^2 = 4a(x - h)$	$(h + a, k)$	$(h + a, k + 2a)$	$(h + a, k - 2a)$	to the right
$(y - k)^2 = -4a(x - h)$	$(h - a, k)$	$(h - a, k + 2a)$	$(h - a, k - 2a)$	to the left
Axis Parallel to Y-axis				
$(x - h)^2 = 4a(y - k)$	$(h, k + a)$	$(h + 2a, k + a)$	$(h - 2a, k + a)$	upward
$(x - h)^2 = -4a(y - k)$	$(h, k - a)$	$(h + 2a, k - a)$	$(h - 2a, k - a)$	downward

Simplify:
$$\left(x - \frac{3}{2}\right)^2 = 8\left(y - \frac{5}{2}\right)$$

$$V\left(\frac{3}{2}, \frac{5}{2}\right), a = 2, F\left(\frac{3}{2}, \frac{9}{2}\right), A\left(\frac{11}{2}, \frac{9}{2}\right), B\left(-\frac{5}{2}, \frac{9}{2}\right)$$

Axis parallel to Y-axis and curve opens upward. The parabola is shown in Figure 29.

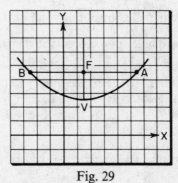

Fig. 29

2) Find the characteristic points of the parabola

$$y^2 - 4x - 6y + 5 = 0$$

Solution.   Transform the equation into the form

$$(y - 3)^2 = 4(x + 1)$$

$$V(-1, 3), a = 1, F(0, 3), A(0, 5), B(0, 1)$$

Axis parallel to X-axis and curve opens to the right. The parabola is shown in Figure 30, p. 124.

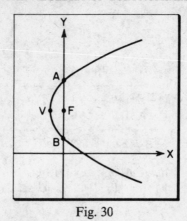

Fig. 30

The Ellipse.   An ellipse is a closed oval. The general quadratic is placed in the standard form

$$\frac{(x - h)^2}{a^2} + \frac{(y - k)^2}{b^2} = 1$$

by completing the squares on both x and y and collecting the terms. The center of the ellipse is at $C(h, k)$. The larger of a and b is the length of the *semimajor axis*, the smaller is the length of the *semiminor axis*. The coordinates of the points at the ends of the major

and minor axes are obtained by adding and subtracting these lengths to the coordinates of the center. These four points are sufficient to sketch the ellipse. If greater accuracy is desired, additional points should be calculated. The graph is symmetric with respect to both the major and minor axes.

    Illustrations. Determine the characteristic points of each of the following ellipses:

1) $9x^2 + 16y^2 - 18x + 96y + 9 = 0$

Solution.   Group the terms to complete the square.

$$9(x^2 - 2x + 1) + 16(y^2 + 6y + 9) = -9 + 9 + 16(9)$$
$$9(x - 1)^2 + 16(y + 3)^2 = 16(9)$$
$$\frac{(x - 1)^2}{16} + \frac{(y + 3)^2}{9} = 1$$

The center is at $(1, -3)$, $a = 4$, and $b = 3$. The end-points of the axes are $A_1(5, -3)$, $A_2(-3, -3)$, $B_1(1, 0)$, $B_2(1, -6)$. The ellipse is shown in Figure 31.

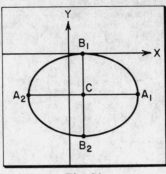

Fig. 31

2) $25x^2 + 16y^2 = 400$

Solution.   A division by 400 yields

$$\frac{x^2}{16} + \frac{y^2}{25} = 1$$

The center is at $C(0, 0)$. The major axis is along the Y-axis with end-points $(0, 5)$ and $(0, -5)$. The minor axis is along the X-axis with end-points $(4, 0)$ and $(-4, 0)$. The ellipse is shown in Figure 32.

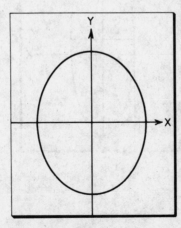

Fig. 32

3) $196x^2 - 588x + 36y^2 - 36y + 9 = 0$

Solution. Complete the squares:

$$196\left(x^2 - 3x + \frac{9}{4}\right) + 36\left(y^2 - y + \frac{1}{4}\right) = -9 + 9(49) + 9$$

$$196\left(x - \frac{3}{2}\right)^2 + 36\left(y - \frac{1}{2}\right)^2 = 9(49)$$

$$\frac{4\left(x - \frac{3}{2}\right)^2}{9} + \frac{4\left(y - \frac{1}{2}\right)^2}{49} = 1$$

$$\frac{\left(x - \frac{3}{2}\right)^2}{\left(\frac{9}{4}\right)} + \frac{\left(y - \frac{1}{2}\right)^2}{\left(\frac{49}{4}\right)} = 1$$

The center is at $(3/2, 1/2)$. Since $a = 3/2$ and $b = 7/2$, the major axis is parallel to the *Y*-axis with end-points $(3/2, 4)$ and $(3/2, -3)$. The end-points of the minor axis are $(3, 1/2)$ and $(0, 1/2)$. The ellipse is shown in Figure 33.

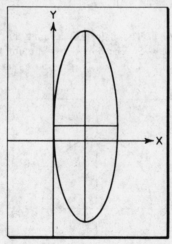

Fig. 33

4) $x^2 + y^2 - 6x - 2y = 15$

Solution. Complete the squares:

$$(x^2 - 6x + 9) + (y^2 - 2y + 1) = 15 + 9 + 1 = 25$$

$$(x - 3)^2 + (y - 1)^2 = 25$$

This is a circle with center at $(3, 1)$ and radius $= 5$; it is shown in

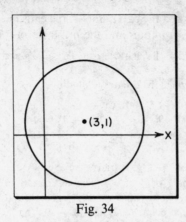

Fig. 34

Figure 34.

The Hyperbola. If A and C have *opposite* signs in the general quadratic with $B = 0$, then the graph is a hyperbola. This graph consists of two branches each *looking like* a parabola. The general quadratic is transformed into one of the standard forms

$$\frac{(x - h)^2}{a^2} - \frac{(y - k)^2}{b^2} = \pm 1$$

If the right member is $+1$, then the *transverse axis* is parallel to the X-axis and the *semitransverse axis* is a; the *conjugate axis* is parallel to the Y-axis and the *semiconjugate axis* is b. If the right member is -1, the above statements relative to the transverse axis and conjugate axis are reversed. The vertices of the hyperbola are on the transverse axis and at a distance equal to the length of the semitransverse on either side of the center. There are two foci located on the transverse axis at a distance of $\pm \sqrt{a^2 + b^2}$ from the center. A line perpendicular to the transverse axis through each focus intersects the hyperbola at two points. The line segment between these points is called the *latus rectum*, L.R., and its length is equal to twice the square of the semiconjugate axis divided by the semitransverse axis. Thus, if the right member is $+1$, then L.R. = $2b^2/a$. The lines through the center and the points $(h + a, k + b)$ and $(h + a, k - b)$ are called the **asymptotes** of the hyperbola. (See Section 71 for a general definition of asymptotes.) The hyperbola does not cross the asymptotes.

To sketch a hyperbola, find its center, the transverse axis, the vertices, the foci, and the two points at the ends of each latus rectum. Then draw the asymptotes and the branches of the hyperbola.

Illustrations.   Find the characteristic parts and sketch the hyperbolas.

1) $9x^2 - 16y^2 - 36x - 32y - 124 = 0$

Solution.   Complete the squares to obtain

$$9(x^2 - 4x + 4) - 16(y^2 + 2y + 1) = 124 + 36 - 16$$
$$9(x - 2)^2 - 16(y + 1)^2 = 144$$
$$\frac{(x - 2)^2}{16} - \frac{(y + 1)^2}{9} = 1$$

The center is at $(2, -1)$, the transverse axis is parallel to the X-axis, $a = 4$, $b = 3$, and $\sqrt{a^2 + b^2} = 5$. The vertices are $(6, -1)$ and $(-2, -1)$. The foci are $(7, -1)$ and $(-3, -1)$. The asymptotes pass through the center and the points $(6, 2)$ and $(6, -4)$. The length of the latus rectum is $2b^2/a = 9/2$. The four points on the hyperbola are $(7, 5/4)$, $(7, -13/4)$, $(-3, 5/4)$, and $(-3, -13/4)$. The graph is shown in Figure 35.

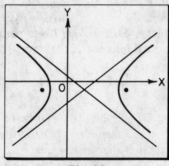

Fig. 35

2) $16x^2 - 9y^2 + 36 = 0$

Solution.   Transform the equation into

$$\frac{x^2}{\frac{9}{4}} - \frac{y^2}{4} = -1$$

The center is at $(0, 0)$, the transverse axis is along the Y-axis, $a = 3/2$, $b = 2$, and $\sqrt{a^2 + b^2} = 5/2$. The vertices are $(0, 2)$ and $(0, -2)$. The foci are $(0, 5/2)$ and $(0, -5/2)$. The asymptotes pass through the origin and $(3/2, 2)$ and $(3/2, -2)$. The length of the latus rectum is $2a^2/b = 9/4$. The hyperbola is shown in Figure 36.

If $a = b$, the hyperbola is called an *equilateral hyperbola* and the asymptotes are perpendicular to each other. The form

$$xy = K$$

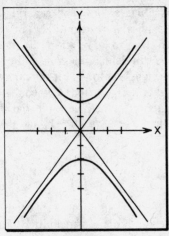

Fig. 36

where K is a constant is an equilateral hyperbola for which the asymptotes are the coordinate axes. The two branches may be plotted directly from a calculated table of ordered pairs.

Illustration.   Sketch the graph of $xy = 8$.
Solution.   Calculate the table of values:

x	± 1	± 2	± 3	± 4	± 6	± 8
y	± 8	± 4	$\pm \dfrac{8}{3}$	± 2	$\pm \dfrac{4}{3}$	± 1

The graph is shown in Figure 37. Notice that the vertices lie on the line $y = x$, and the coordinates are $x = y = \pm \sqrt{8}$.

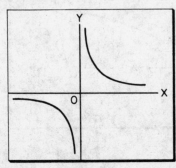

Fig. 37

66. The Inverse of Quadratics. The general quadratic equation in two variables defines a relation. The inverse of the general quadratic equation is also a relation. By limiting the domain and/or the range we can define functions and obtain their inverses. We shall limit our discussion to the set of real numbers.

Consider the quadratic function defined by the equation $y = 2x^2 - 4x$. This function is defined over the domain of all real numbers. To find the inverse we interchange x and y and solve for y to obtain

$$y = 1 \pm \sqrt{1 + \frac{x}{2}}$$

This, however, is a relation, since for each x there are two values of y. Let us now limit the domain of definition of the given function to be the interval $[1, +\infty[$. The range is then the interval $[-2, +\infty[$. Recall that in our definition of the inverse the range of f becomes the domain of f^{-1} and the domain of f becomes the range of f^{-1}. Since the range of f^{-1}, $[1, +\infty[$, is positive for all x we choose the plus sign and write

$$f^{-1}: y = 1 + \sqrt{1 + \frac{x}{2}}$$

thus obtaining an inverse function. The graphs are shown in Figure 38 with the restricted function and its inverse shown by heavy arcs. We shall repeat the procedure for an ellipse.

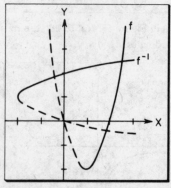

Fig. 38

Illustration. Find the inverse function of f defined by

$$y = \frac{1}{2}\sqrt{16 - x^2}$$

over the domain $-4 \le x \le 0$.

Solution. Interchange the variables to obtain

$$x = \frac{1}{2}\sqrt{16 - y^2}$$

Square both sides and simplify:

$$x^2 = \frac{1}{4}(16 - y^2)$$
$$y^2 = 16 - 4x^2$$

Extract the square root of both sides to obtain

$$y = \pm 2\sqrt{4 - x^2}$$

We are now faced with the problem of choosing the proper sign. Since the domain of f is the set $[-4, 0]$, the range of f^{-1} must be negative, so the inverse of f is defined by

$$y = -2\sqrt{4 - x^2}$$

The two functions are shown in Figure 39.

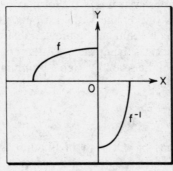

Fig. 39

67. Exercise VII.

1. Find the solutions of the following quadratic equations:

a) $2x^2 - 4x = 0$ b) $x^2 - 17 = 0$

c) $x^2 - x - 6 = 0$ d) $4x^2 + 2x - 15 = 0$

e) $x^2 + x + 1 = 0$ f) $2x^2 + 3x - 4 = 0$

g) $9x^2 + 12x + 4 = 0$ h) $4x^2 - 4x - 11 = 0$

i) $3x^2 - 6x + 5 = 0$ j) $x^2 + 2x + 3 = 0$

2. Solve the linear equation for y in terms of x and substitute into the quadratic equation; then solve the resulting quadratic for x. Substitute back into the linear equation to find the corresponding values of y, thus finding the solutions for the following systems:

a) $\begin{cases} 2x + 3y = 6 \\ 3y^2 + xy = 0 \end{cases}$
   b) $\begin{cases} x + y = 8 \\ x^2 - 4x + 4 = 6y - y^2 \end{cases}$

3. Solve the equation $bxy + f = 0$ for y or x and substitute into $ax^2 + cy^2 + f = 0$. The result is a quadratic in the square of the unknown which can be solved. The corresponding value of the other unknown is then obtained from the first equation. Solve the following systems in this manner:

a) $\begin{cases} xy = 4 \\ x^2 + 4y^2 = 20 \end{cases}$
   b) $\begin{cases} 2xy = 3 \\ 36x^2 + 16y^2 = 97 \end{cases}$

4. Treat the following systems as linear in x^2 and y^2 and solve as systems of linear equations:

a) $\begin{cases} 4x^2 + 5y^2 = 16 \\ 13y^2 - 5x^2 = 57 \end{cases}$
   b) $\begin{cases} x^2 - 4y^2 = 32 \\ 2y^2 - 5x^2 = 2 \end{cases}$

5. If one equation of a system is of the form $ax^2 + by^2 + f = 0$ and the other contains one of the variables only as a squared term, the system is solved by eliminating the variable which is quadratic in both equations and solving the resulting quadratic in one unknown. The value of the second variable is obtained by substituting into either equation.

a) $\begin{cases} x^2 + 4y^2 - 25 = 0 \\ x^2 - 2y - 5 = 0 \end{cases}$
   b) $\begin{cases} x^2 - 4y^2 = 5 \\ x^2 - 10y = 19 \end{cases}$

6. If both equations of a system of quadratics are of the form $ax^2 + bxy + cy^2 + f = 0$ (i.e., they contain no linear terms), eliminate the constant term f and factor the resulting equation into two linear equations. The system can then be solved by the method of Exercise 2, above. Use this hint to solve the following systems:

a) $\begin{cases} 3x^2 + xy + 2y^2 - 6 = 0 \\ 3x^2 + xy + 4y^2 - 9 = 0 \end{cases}$

b) $\begin{cases} 4x^2 + 9y^2 - 10 = 0 \\ 2x^2 - 3xy - 2 = 0 \end{cases}$

c) $\begin{cases} 3x^2 + 2xy - 4y^2 - 9 = 0 \\ 2x^2 + 3xy + y^2 - 14 = 0 \end{cases}$

7. If both equations of a system of quadratics are symmetric $[A(x^2 + y^2) + Bxy + D(x + y) + F = 0]$ make the substitutions $x = u + v$ and $y = u - v$; then $x^2 + y^2 = 2(u^2 + v^2)$, $xy = u^2 - v^2$, and $x + y = 2u$. The resulting equations will have no linear term in v. Eliminate the v^2 term and solve the resulting quadratic in u. A substitution will obtain v and subsequently x and y. Solve the following systems:

a) $\begin{cases} x^2 + 3xy + y^2 + 2x + 2y - 5 = 0 \\ 3x^2 - 2xy + 3y^2 - 5x - 5y - 26 = 0 \end{cases}$

b) $\begin{cases} x^2 + y^2 + 2(x + y) - 11 = 0 \\ 3xy - 2(x + y) = 0 \end{cases}$

8. Sketch the graphs of the following quadratics:

a) $3x^2 - 6x + y - 3 = 0$ b) $2y^2 = x$

c) $x^2 + y^2 - 16 = 0$ d) $2x^2 = y$

e) $4y^2 - 16x - 20y - 7 = 0$ f) $xy = 4$

g) $y = 2x^2 - 3x + 5$ h) $2xy + 5 = 0$

i) $x^2 - y^2 = 1$ j) $4x^2 + 9y^2 = 36$

k) $25x^2 + 16y^2 + 100x - 32y - 284 = 0$

l) $16x^2 - 48x + 144 = 9y^2 - 36y$

m) $4x^2 + y^2 + 24x - 16y + 84 = 0$

9. Find the inverse of each of the following functions:

a) $f : x^2 - 4y = 0, \quad d_f : 0 \le x \le 4$

b) $f : y = \sqrt{16 - x^2}, \quad d_f : -4 \le x \le 0$

c) $f : y = -\dfrac{2}{3}\sqrt{9 - x^2}, \quad d_f : [-3, 0]$

d) $f : y = \sqrt{x^2 - 9}, \quad d_f : [3, 5]$

68. Polynomials of Higher Degree. We now turn our attention to polynomials of higher degree than two. Let us denote the general polynomial in one variable by

$$P(x) = a_0 x^n + a_1 x^{n-1} + a_2 x^{n-2} + \ldots + a_{n-1} x + a_n$$

where n is a positive integer and a_i $(i = 0, 1, \ldots, n)$ are constants with $a_0 \ne 0$. The polynomial can be defined with x and the coefficients a_i as members of the set of complex numbers; however, we shall limit the coefficients to the real numbers. The function defined by the above equation is a **rational integral function of the nth degree in x.** The corresponding rational integral equation is obtained by setting the expression equal to 0.

Illustration.　Arrange the following equation in standard form and give the values of n and a_i:

$$(3x^2 - 2)^2 + kx^2 = 4x - 1$$

Solution.　Perform the indicated multiplication and simplify:

$$9x^4 - 12x^2 + 4 + kx^2 = 4x - 1$$
$$9x^4 - 12x^2 + kx^2 + 4 - 4x + 1 = 0$$
$$9x^4 + (k - 12)x^2 - 4x + 5 = 0$$

$$n = 4, \quad a_0 = 9, \quad a_1 = 0, \quad a_2 = k - 12, \quad a_3 = -4, \quad a_4 = 5$$

Remainder Theorem.　*If a polynomial* $P(x)$ *is divided by* $(x - r)$ *until a remainder independent of* x *is obtained, this remainder is equal to* $P(r)$.

Proof.　By definition we have

$$\text{Dividend} = \text{divisor} \cdot \text{quotient} + \text{remainder}$$
$$P(x) = (x - r) Q(x) + R$$

If the polynomial $P(x)$ is of degree n, then $Q(x)$ is of degree $(n - 1)$ in x and R is a constant. The definition is true for all values of x and, in particular, for $x = r$; thus

$$P(r) = (r - r) Q(r) + R$$

Since $Q(r)$ is a number

$$(r - r) Q(r) = 0 \cdot Q(r) = 0$$

and we have

$$P(r) = R$$

Factor Theorem.　*If* r *is a zero of the polynomial* $P(x)$, *then* $(x - r)$ *is a factor of* $P(x)$.

Proof.　Divide $P(x)$ by $(x - r)$ and use the Remainder Theorem to write

$$P(x) = (x - r) Q(x) + P(r)$$

By hypothesis, r is a zero of $P(x)$; i.e., $P(r) = 0$.

$\therefore$　　　　　$$P(x) = (x - r) Q(x)$$

and $(x - r)$ is a factor of $P(x)$.

Illustrations.

1) Find the remainder when $x^4 + x^3 - 2x^2 + x + 3$ is divided by $(x + 2)$.

Solution. Since we are dividing by $x + 2$ we have $r = -2$ and by the Remainder Theorem

$$R = P(r) = P(-2)$$
$$= (-2)^4 + (-2)^3 - 2(-2)^2 + (-2) + 3$$
$$= 1$$

2) Determine whether $(x - 2)$ if a factor of the polynomial

$$2x^4 - x^3 - 13x^2 + 15x - 2.$$

Solution. Since $r = 2$ and

$$P(2) = 2(2)^4 - (2)^3 - 13(2)^2 + 15(2) - 2$$
$$= 32 - 8 - 52 + 30 - 2$$
$$= 62 - 62 = 0$$

$x - 2$ is a factor by the Factor Theorem.

3) Determine a_2 and a_3 so that $x^4 - 3x^3 + a_2x^2 + a_3x - 18$ is divisible by $(x - 2)(x - 3)$.

Solution. By the Factor Theorem $P(2)$ and $P(3)$ must be 0.

$$P(2) = 16 - 24 + 4a_2 + 2a_3 - 18 = 4a_2 + 2a_3 - 26 = 0$$
$$P(3) = 81 - 81 + 9a_2 + 3a_3 - 18 = 9a_2 + 3a_3 - 18 = 0$$

The solution of this system of linear equations is $a_2 = -7$ and $a_3 = 27$.

69. Synthetic Division. The process of dividing a polynomial by a binomial is shortened by a method called **synthetic division**, which we shall now explain.

a) Arrange the polynomial $P(x)$ in descending powers of x.

$$P(x) = a_0x^n + a_1x^{n-1} + a_2x^{n-2} + \ldots + a_{n-1}x + a_n$$

b) Write the coefficients a_i in order on a line. Supply a zero for each missing term.

c) To divide by $(x - r)$ write r at the right on the first line.

d) Complete the following array:

a_0	a_1	a_2	a_3	$\ldots$	a_{n-1}	a_n	r
	b_0r	b_1r	b_2r	$\ldots$	$b_{n-2}r$	$b_{n-1}r$	
b_0	b_1	b_2	b_3	$\ldots$	b_{n-1}	b_n	

where

$$b_0 = a_0$$

$$b_1 = a_1 + b_0 r$$
$$b_2 = a_2 + b_1 r$$
$$. \quad . \quad . \quad . \quad . \quad . \quad . \quad .$$
$$b_i = a_i + b_{i-1} r$$
$$. \quad . \quad . \quad . \quad . \quad . \quad . \quad .$$
$$b_n = a_n + b_{n-1} r = P(r) = R$$

Note that we *add* the elements of the second line to those of the first line to obtain the third line. The set b_i ($i = 0, \ldots, n - 1$) forms the coefficients of the quotient, which is of degree $(n - 1)$.

Illustrations.

1) Divide $2x^4 - 3x^3 + x^2 - x + 2$ by $x - 2$.
Solution.

$$
\begin{array}{rrrrr|l}
2 & -3 & 1 & -1 & 2 & \;2 \\
  & 4 & 2 & 6 & 10 & \\
\hline
2 & 1 & 3 & 5 & 12 &
\end{array}
$$

$$Q(x) = 2x^3 + x^2 + 3x + 5, \quad R = 12$$

2) Divide $x^4 - 3x^3 - 7x^2 + 27x - 18$ by $x + 3$.
Solution.

$$
\begin{array}{rrrrr|l}
1 & -3 & -7 & 27 & -18 & \;-3 \\
  & -3 & 18 & -33 & 18 & \\
\hline
1 & -6 & 11 & -6 & 0 &
\end{array}
$$

The quotient is $x^3 - 6x^2 + 11x - 6$; since the remainder is 0, $x + 3$ is a factor of $P(x)$.

70. Solution Set of Polynomial Equations. The rational integral equation

$$P(x) = 0$$

can be solved to obtain a set of numbers which satisfy the equation and belong to the set of complex numbers. In this section we shall consider some of the interesting characteristics of this solution set.

Theorem 1. *Every rational integral equation* $f(x) = 0$ *has at least one root.*

This is known as the fundamental theorem of algebra. We shall not give the proof of this theorem. Its proof involves the theory of complex variables, which is beyond the scope of this book.*

---

*See R. V. Churchill, *Complex Variables and Applications,* New York: McGraw-Hill Book Company, Inc., 1960, p. 125.

Theorem 2. *Every rational integral equation of the nth degree has n roots and no more.*

Proof. Let

$$P(x) = a_0x^n + a_1x^{n-1} + \ldots + a_{n-1}x + a_n = 0, a_0 \neq 0$$

By Theorem 1, there is at least one root; call it r_1. Then by the Factor Theorem we have

$$P(x) = (x - r_1)\, Q_1(x)$$

where $Q_1(x)$ is a polynomial of degree $n - 1$. The equation $Q_1(x) = 0$ has at least one root, r_2, and by the Factor Theorem we have

$$Q_1(x) = (x - r_2)\, Q_2(x)$$

where $Q_2(x)$ is a polynomial of degree $n - 2$. Continue this process n times to obtain

$$P(x) = (x - r_1)\, (x - r_2)\, (x - r_3) \ldots (x - r_n)\, Q_n(x)$$

where

$$Q_n(x) = a_0x^{n-n} = a_0$$

Therefore

$$P(x) = a_0(x - r_1)\, (x - r_2) \ldots (x - r_n) = 0$$

and we have exactly n roots, $r_1, r_2, \ldots, r_n$. It remains to demonstrate that there are no more roots and we shall do this by a contradiction. Suppose that there is a root r different from r_i, $(i = 1, \ldots, n)$; then in

$$P(r) = a_0(r - r_1)\, (r - r_2) \ldots (r - r_n)$$

none of the factors $(r - r_i)$ is 0, and since $a_0 \neq 0$ we have $P(r) \neq 0$, so that r cannot be a root of $P(x) = 0$. This completes the proof of the theorem.

We shall now list some characteristics of the solution set $S = \{x_i, i = 1, \ldots, n\}$ of the polynomial equation

$$P(x) = a_0x^n + a_1x^{n-1} + \ldots + a_{n-1}x + a_n = 0$$

with real coefficients.

    I. *Nonreal roots.* If $x_1 = a + bi \in S$, then $x_2 = a - bi \in S$.

    II. *Quadratic surds.* If $a + \sqrt{b} \in S$, then $a - \sqrt{b} \in S$.

    III. *Descartes' Rule of Signs.* The number of *positive* roots, $x_i > 0$, cannot exceed the number of variations in sign in $P(x)$; the number of *negative* roots, $x_i < 0$, cannot exceed the number of variations in sign in $P(-x)$. (A variation in sign occurs whenever two successive terms in $P(x)$ differ in sign. To find $P(-x)$ simply change the sign of the odd-power terms.)

IV. *Odd and even degree.* If n is odd, the equation has at least one real root; if n is even and x_1 is real, then there is at least one more real root.

V. *Rational roots.* If the coefficients a_i are integers and $x_i = b/c$ is a rational number in its lowest terms, then b is a factor of a_n and c is a factor of a_0.

VI. *Sums and products of roots.*

$$\sum_{i=1}^{n} x_i = x_1 + x_2 + x_3 + \ldots + x_n$$

See Section 85. If $x_1 \in S$, then

$$\sum_{i=1}^{n} x_i = -\frac{a_1}{a_0}$$

$$\sum_{i,j=1}^{n} x_i x_j = \frac{a_2}{a_0} \qquad (i \neq j)$$

$$\sum_{1}^{n} x_i x_j x_k = -\frac{a_3}{a_0} \qquad (i \neq j \neq k)$$

$$x_1 \cdot x_2 \ldots x_n = (-1)^n \frac{a_n}{a_0}$$

To find the solution set for a given equation we use the above characteristics in the following manner:

a) Find the maximum number of positive and negative roots.

b) Find the rational roots and depress the equation by factoring them out.

c) If the resulting reduced equation is quadratic, solve it by use of the quadratic formula.

Illustrations. Solve the following equations:

1) $2x^4 - x^3 - 14x^2 - 5x + 6 = 0$

Solution.

$P(x): \quad + - - - +$ (2 variations)

$P(-x): + + - + +$ (2 variations)

Factors of $a_n = 6: \pm 1, \pm 2, \pm 3, \pm 6$

Factors of $a_0 = 2: \pm 1, \pm 2$

$$\therefore \frac{b}{c} = \pm 1, \pm 2, \pm 3, \pm 6, \pm \frac{1}{2}, \pm \frac{3}{2}$$

Use synthetic division to find $P(x) = 0$:

$$
\begin{array}{rrrrr|l}
2 & -1 & -14 & -5 & 6 & \underline{\hspace{0.3cm}-1} \\
  & -2 & 3 & 11 & -6 & \\
\hline
2 & -3 & -11 & 6 & 0 &
\end{array}
\qquad \therefore x = -1
$$

$$
\begin{array}{rrrr|l}
 & -4 & 14 & -6 & \underline{\hspace{0.3cm}-2} \\
\hline
2 & -7 & 3 & 0 &
\end{array}
\qquad \therefore x = -2
$$

$$
\begin{array}{rrr|l}
 & 6 & -3 & \underline{\hspace{0.3cm}3} \\
\hline
2 & -1 & 0 &
\end{array}
\qquad \therefore x = 3
$$

The equation has been reduced to

$$2x - 1 = 0 \qquad \text{and} \qquad x = \frac{1}{2}$$

$$\therefore \qquad\qquad S = \left\{ -1, -2, 3, \frac{1}{2} \right\}$$

2) $2x^5 + 10x^4 + 11x^3 - 4x^2 - x + 6 = 0$

Solution.

No more than 2 positive roots

No more than 3 negative roots

Factors of $a_n = 6: \pm 1, \pm 2, \pm 3, \pm 6$

Factors of $a_0 = 2: \pm 1, \pm 2$

$$\therefore \frac{b}{c}: \pm 1, \pm 2, \pm 3, \pm 6, \pm \frac{1}{2}, \pm \frac{3}{2}$$

By trial:

$$
\begin{array}{rrrrrr|l}
2 & 10 & 11 & -4 & -1 & 6 & \underline{\hspace{0.3cm}-1} \\
  & -2 & -8 & -3 & 7 & -6 & \\
\hline
2 & 8 & 3 & -7 & 6 & 0 &
\end{array}
$$

$$
\begin{array}{rrrrr|l}
 & -4 & -8 & 10 & -6 & \underline{\hspace{0.3cm}-2} \\
\hline
2 & 4 & -5 & 3 & 0 &
\end{array}
$$

$$
\begin{array}{rrrr|l}
 & -6 & 6 & -3 & \underline{\hspace{0.3cm}-3} \\
\hline
2 & -2 & 1 & 0 &
\end{array}
$$

The depressed equation is $2x^2 - 2x + 1 = 0$, which is solved by the quadratic formula to give

$$x = \frac{1}{2} \pm \frac{1}{2}i$$

$$\therefore \qquad S = \left\{ -1, -2, -3, \frac{1}{2} + \frac{1}{2}i, \frac{1}{2} - \frac{1}{2}i \right\}$$

71. Algebraic Curves. In Section 65 we discussed the graphs of the quadratics in some detail. Since we are drawing the graphs in the real plane, we noticed that the domain of definition for the conics was limited. Thus in order for the function defined by

$$y = \frac{1}{2}\sqrt{x^2 - 9}$$

to be real, x cannot be in the open interval $]-3, 3[$. The graph of this function is the two positive halves of the hyperbola, thus the range is $0 \leq y < \infty$.

The above function is an example of an *explicit algebraic function* since it is solved for y explicitly; i.e., it is defined in the form $y = f(x)$. We shall now consider some algebraic functions of this type.

The Polynomial Function. A polynomial function defined by $y = P(x)$ has the entire set of real numbers for its domain and some subset of the real numbers for its range. We can obtain its graph by calculating a table of values, plotting the points, and joining them with a smooth curve. In graphing polynomial functions, the scale of the Y-axis may be different from that of the X-axis if desired.

Illustration.   Sketch the graph of the function defined by $y = x^3 - 4x$.
Solution.   Calculate a table of values:

x	-3	-2	-1	0	1	2	3
y	-15	0	3	0	-3	0	15

The interesting part of the curve is the portion defined by x in the interval $[-2, 2]$. We therefore calculate a few more values in this interval.

x	$-\dfrac{3}{2}$	$-\dfrac{1}{2}$	$\dfrac{1}{2}$	$\dfrac{3}{2}$
y	$\dfrac{21}{8}$	$\dfrac{15}{8}$	$-\dfrac{15}{8}$	$-\dfrac{21}{8}$

The curve is shown in Figure 40.

An interesting aspect of the graph of a polynomial is the location of the top (or bottom) of the bends. These are called the *relative maxima and minima* and constitute part of the study of curve-tracing in the calculus. We shall indicate a procedure for finding these

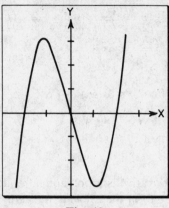

Fig. 40

points without lingering on the rigorous proofs.

In Section 55 we discussed the rate of change of a linear function and developed the idea of finding the change of a function through an incremental change of the argument. Recall that

$$y + \Delta y = f(x + \Delta x)$$

and that

$$\Delta y = f(x + \Delta x) - f(x)$$

Let us divide this change in y by the increment in x to form the fraction

$$\frac{\Delta y}{\Delta x} = \frac{f(x + \Delta x) - f(x)}{\Delta x}$$

For a polynomial of degree higher than one, the right-hand side will contain Δx; consequently this fraction depends upon the increment Δx. But now suppose that we let Δx become smaller and smaller and finally let $\Delta x \to 0$. The fraction would then take on an undefined form, since Δx occurs in the denominator. In the calculus we define this limit to be the **derivative** of y with respect to x. It is easy to see that for a polynomial the expression will exist by considering the right-hand side.

Illustration.   Find the derivative of y with respect to x if $y = x^3 - x + 2$.
Solution.   Form the fraction $\Delta y/\Delta x$.

$$\frac{\Delta y}{\Delta x} = \frac{f(x + \Delta x) - f(x)}{\Delta x}$$

$$= \frac{1}{\Delta x}[(x + \Delta x)^3 - (x + \Delta x) + 2 - (x^3 - x + 2)]$$

$$= \frac{1}{\Delta x}[x^3 + 3x^2\Delta x + 3x(\Delta x)^2 + (\Delta x)^3 - x - \Delta x$$

$$+ 2 - x^3 + x - 2]$$

$$= \frac{1}{\Delta x}[3x^2\Delta x + 3x(\Delta x)^2 + (\Delta x)^3 - \Delta x]$$

$$= 3x^2 + 3x\Delta x + (\Delta x)^2 - 1$$

Now let $\Delta x = 0$; then

$$D_x y = 3x^2 - 1$$

Notice that the derivative is a polynomial of degree one less than the given polynomial. In the calculus we learn:

$$\text{If } y = ax^n, \text{ then } D_x y = anx^{n-1},$$
$$\text{if } y = c, \text{ then } D_x y = 0,$$

and

$$D_x(y_1 + y_2) = D_x y_1 + D_x y_2$$

Assuming the truth of these two formulas we can write the derivatives of polynomials.

Examples:
a) If $y = 2x^3 - 3x^2 + x$, then $D_x y = 6x^2 - 6x + 1$.
b) If $y = x^2 - 3x - 1$, then $D_x y = 2x - 3$.
c) If $y = x^5 - 6x^4 + x^3$, then $D_x y = 5x^4 - 24x^3 + 3x^2$.

We shall further assume the following property:

Property. *The relative maxima and minima of a polynomial occur at those values of the argument for which the derivative is zero.*

Illustration. Find the relative extrema (maxima and minima) of

$$y = 2x^3 - 3x^2 - 12x + 6$$

Solution. Find the derivative:

$$D_x y = 6x^2 - 6x - 12$$

Set the derivative equal to zero and solve:

$$6x^2 - 6x - 12 = 0$$
$$x^2 - x - 2 = 0$$
$$(x - 2)(x + 1) = 0$$
$$x = 2, -1$$

The extrema are

$$y_1 = f(2) = 2(8) - 3(4) - 12(2) + 6 = -14$$
$$y_2 = f(-1) = 2(-1) - 3(1) - 12(-1) + 6 = 13$$

Whenever the zeros of the derivative polynomial are easily found, we shall use them to find characteristic points in plotting graphs.

Illustration.   Plot the graph of $y = x^3 - 2x^2$.
Solution.   Find the derivative:

$$D_x y = 3x^2 - 4x$$

The zeros of this polynomial are $x = 0$ and $x = 4/3$. The zeros of the given polynomial are $x = 0$ (twice) and $x = 2$. Calculate a table of values for these and adjacent values of x.

x	-1	0	1	$\dfrac{4}{3}$	2	3
y	-3	0	-1	$-\dfrac{32}{27}$	0	9

The graph is shown in Figure 41. Notice that there is a relative maximum at $x = 0$ and a minimum at $x = 4/3$.

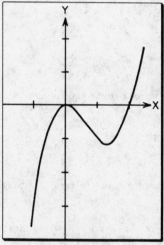

Fig. 41

The Rational Function.   A function defined by

$$y = \frac{P(x)}{Q(x)}$$

where $P(x)$ and $Q(x)$ are polynomials is called a **rational function.** It is defined for all values of x except at the zeros of $Q(x)$, i.e., the values of x for which $Q(x) = 0$. Although these values of x must be excluded from the domain of definition they define an interesting characteristic of the graph.

Definition. *A line* $x = c$ *is said to be a* **vertical asymptote** *of the graph of a function defined by* $y = f(x)$ *if and only if as* x *approaches* c, *the value of* y *becomes indefinitely large* (*either positively or negatively*) *and the curve gets closer and closer to the line as* y *gets larger and larger.*

If $Q(c) = 0$, then $x = c$ is a vertical asymptote of the curve of $y = P(x)/Q(x)$. Suppose we solve the equation for x in terms of y to obtain

$$x = \frac{R(y)}{S(y)}$$

Then the line $y = k$ is said to be a **horizontal asymptote** of the curve defined by

$$y = \frac{P(x)}{Q(x)} \quad \text{if} \quad S(k) = 0$$

Illustrations.

1) Sketch the graph of

$$y = \frac{x}{x - 1}$$

Solution. Set the denominator equal to zero to obtain the vertical asymptote

$$x = 1$$

Solve for x to obtain

$$x = \frac{y}{y - 1}$$

and set the denominator equal to zero. We obtain the horizontal asymptote

$$y = 1$$

Calculate the table of values:

x	-2	-1	0	$\frac{1}{2}$	$\frac{2}{3}$	1	$\frac{3}{2}$	2	3
y	$\frac{2}{3}$	$\frac{1}{2}$	0	-1	-2	$-$	3	2	$\frac{3}{2}$

The graph is shown in Figure 42.

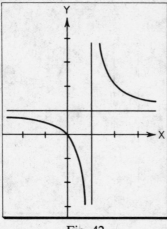

Fig. 42

2) Sketch the graph of

$$y = \frac{x}{(x + 1)(x - 2)}$$

Solution. The vertical asymptotes are

$$x = -1 \qquad \text{and} \qquad x = 2$$

Solve for x to obtain

$$x = \frac{1 + y \pm \sqrt{9y^2 + 2y + 1}}{2y}$$

Although $y = 0$ is a horizontal asymptote for certain branches of the curve, we note that at $y = 0$ we have

$$x = \frac{1 + 1}{0} = \frac{2}{0} \text{ and } \frac{0}{0}$$

The second of these is an indeterminate form and we return to the original equation to see that the point $(0, 0)$ is on the curve. Calculate the following table of values:

x	-3	-2	$-\dfrac{3}{2}$	-1	$-\dfrac{1}{2}$	0	$\dfrac{1}{2}$	1	$\dfrac{3}{2}$	2	$\dfrac{5}{2}$	3	4
y	$-\dfrac{3}{10}$	$-\dfrac{1}{2}$	$-\dfrac{6}{7}$		$\dfrac{2}{5}$	0	$-\dfrac{2}{9}$	$-\dfrac{1}{2}$	$-\dfrac{6}{5}$		$\dfrac{10}{7}$	$\dfrac{3}{4}$	$\dfrac{2}{5}$

The graph is shown in Figure 43.

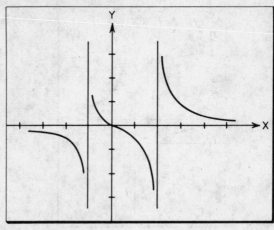

Fig. 43

Since the value of a polynomial function is 0 at the points where the graph crosses the X-axis, we can approximate the real roots of the equation $P(x) = 0$.

Illustration. Find the real roots of

$$x^3 - x^2 - x + 3 = 0$$

Solution. Calculate the table of values:

x	-2	-1	0	1	2
y	-7	2	3	2	5

and sketch the graph. See Figure 44 (p. 147). The curve crosses the X-axis between -2 and -1. (The other two roots are not real.) It will be closer to -1 than -2 since $|y = 2| < |y = -7|$. Use synthetic division to calculate the values of $f(x)$ for $x = -1.2, -1.3, -1.4$, etc., until $f(x)$ changes sign.

x	-1.2	-1.3	-1.4
y	1.032	$.413$	$-.304$

The root then is $-1.3 < x_1 < -1.4$. Locate the last two points (x, y) on an enlarged scale and draw a straight line between them. See

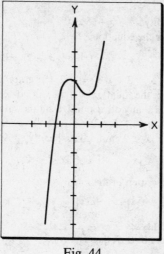

Fig. 44

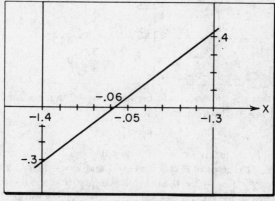

Fig. 45

Figure 45. The root is approximately $x = -1.36$.

72. Polynomial Inequalities. We discussed linear inequalities in Section 59 and shall now consider the general polynomial inequalities. We seek to find the set of numbers for which $P(x) > 0$ or $P(x) < 0$. For the general polynomial the desired set may be the union of a number of subsets. These subsets are readily obtained by considering the graph of $y = P(x)$ and noting that $y > 0$ above the X-axis, $y = 0$ on the X-axis, and $y < 0$ below the X-axis. For this purpose it is not necessary to draw a detailed graph.

Illustrations.

1) Solve the inequality $x^2 - x - 6 > 0$.

Solution. Consider the equation

$$x^2 - x - 6 = (x - 3)(x + 2) = 0$$

Since the roots are $x = 3$ and $x = -2$, the graph crosses the X-axis at these points. Since the equation is of the second degree, the graph will not cross the X-axis at any other point. Find the value of $f(x)$ at a point between -2 and 3, say $x = 0, f(0) = -6 < 0$. The function is therefore negative for $-2 < x < 3$ and, consequently, positive for the set

$$\{x | x > 3\} \cup \{x | x < -2\}$$

which can also be written in the form

$$]-\infty, -2[\cup]3, \infty[$$

2) Solve the inequality $x^3 + x^2 - 2x \geq 0$.

Solution. The roots of the equation are $x = 0, 1, -2$. Calculate the table:

x	-2	-1	0	$\dfrac{1}{2}$	1	2
y	0	2	0	$-\dfrac{5}{8}$	0	8

We see that $y > 0$ for values of x between $x = -2$ and $x = 0$ and to the right of $x = 1$. The solution set is

$$[-2, 0] \cup [1, \infty[$$

3) Solve the inequality $3x^2 + 11x + 8 \leq 0$.

Solution. The zeros of $P(x)$ are $x = -1$ and $x = -8/3$. At $x = -2$ we have $P(-2) = -2 < 0$. Since this is a quadratic and there are no more zeros of the polynomial, we have the inequality satisfied for all x in the interval $[-8/3, -1]$ or $-8/3 \leq x \leq -1$.

4) Show that the sum of any positive number and its reciprocal is not less than 2.

Solution. Let n be a positive number; then we must show that

$$n + \frac{1}{n} \geq 2, \quad n > 0$$

If this relation is true, then by the Properties of Section 33, we can perform the following operations:

Multiply by n: $n^2 + 1 \geq 2n$
Subtract $2n$: $n^2 - 2n + 1 \geq 0$

Factor: $(n - 1)^2 \geq 0$

This statement is true (see Illustration 3, page 53) and we can reverse the steps to arrive at the conclusion.

73. Partial Fractions. Two polynomials of the same degree are identically equal if and only if the coefficients of like-powered terms are equal. Consider the two polynomials

$$P(x) = a_0 x^n + a_1 x^{n-1} + \ldots + a_{n-1}x + a_n$$

and

$$Q(x) = b_0 x^n + b_1 x^{n-1} + \ldots + b_{n-1}x + b_n$$

The above statement says that if $P(x) = Q(x)$ for all x then $a_i = b_i$ for $i = 0, \ldots, n$. We can demonstrate the proposition in the following way. Since the equality is to hold for *all* x, let $x = 0$; then

$$P(0) = Q(0) \text{ yields } a_n = b_n$$

and these two coefficients can be removed from consideration. Factor the remaining portion of the polynomials and write

$$x(a_0 x^{n-1} + a_1 x^{n-2} + \ldots + a_{n-1})$$
$$= x(b_0 x^{n-1} + b_1 x^{n-2} + \ldots + b_{n-1})$$

If this is to be an identity for all x, the two expressions in the parentheses must be equal for all x. If we let $x = 0$ in the parentheses, we obtain $a_{n-1} = b_{n-1}$ and we can remove these terms. The process is continued until we finally arrive at $a_0 = b_0$.

This proposition forms the basis for a mathematical technique known as the *method of undetermined coefficients*. We shall illustrate the method by decomposing a polynomial fraction into the algebraic sum of fractions whose denominators are of lower degree than that of the given fraction. This is the inverse process of combining two fractions by means of a lowest common denominator. (See Section 9.)

The first step is to reduce (if necessary) the given fraction into a *proper* fraction, i.e., one in which the degree of the numerator is less than the degree of the denominator.

Example:

$$\frac{2x^4 - 2x^2 + 5x}{x^2 - 1} = 2x^2 + \frac{5x}{x^2 - 1}$$

We shall limit our discussion to proper fractions whose denominators can be factored into linear and quadratic factors. Consider the following four cases:

Case	Form Functions
I. Different linear factors $(x - a), (x - b)$	$\dfrac{A}{x - a} + \dfrac{B}{x - b}$
II. Repeated linear factors $(x - a)^p$	$\dfrac{A}{x - a} + \dfrac{B}{(x - a)^2} + \ldots + \dfrac{M}{(x - a)^p}$
III. Different nonreducible quadratic factors	$\dfrac{Ax + B}{x^2 + mx + n}$
IV. Repeated nonreducible quadratic factors	$\dfrac{Ax + B}{x^2 + mx + n} + \dfrac{Cx + D}{(x^2 + mx + n)^2} + \ldots$

These fractions are then combined and the undetermined coefficients $A, B, C, \ldots$ are found by equating coefficients of like powers in the numerators.

Illustrations. Resolve the given fractions into their simplest partial fractions.

1) $\dfrac{6x - 4}{x^2 - 1}$

Solution.

$$\frac{6x - 4}{x^2 - 1} = \frac{A}{x - 1} + \frac{B}{x + 1}$$

$$= \frac{A(x + 1) + B(x - 1)}{(x - 1)(x + 1)}$$

$\therefore \quad 6x - 4 = Ax + A + Bx - B = (A + B)x + (A - B)$

We then have $A + B = 6$ and $A - B = -4$. The solution set of this system of linear equations is $A = 1$ and $B = 5$.

$\therefore \quad \dfrac{6x - 4}{x^2 - 1} = \dfrac{1}{x - 1} + \dfrac{5}{x + 1}$

2) $\dfrac{6x^3 - 4x + 1}{x^2(x + 1)^2}$

Solution.

$$\frac{6x^3 - 4x + 1}{x^2(x + 1)^2} = \frac{A}{x} + \frac{B}{x^2} + \frac{C}{x + 1} + \frac{D}{(x + 1)^2}$$

$$= \frac{Ax(x+1)^2 + B(x+1)^2 + Cx^2(x+1) + Dx^2}{x^2(x+1)^2}$$

Set the numerators equal:

$$6x^3 - 4x + 1 = Ax^3 + 2Ax^2 + Ax + Bx^2 + 2Bx + B + Cx^3 + Cx^2 + Dx^2$$
$$= (A+C)x^3 + (2A+B+C+D)x^2 + (A+2B)x + B$$

Since the coefficients of x^j are equal, we have the system

$$\begin{cases} A + C = 6 \\ 2A + B + C + D = 0 \\ A + 2B = -4 \\ B = 1 \end{cases}$$

The solution set is

$$B = 1, A = -6, C = 12, D = -1$$

and

$$\frac{6x^3 - 4x + 1}{x^2(x+1)^2} = -\frac{6}{x} + \frac{1}{x^2} + \frac{12}{x+1} - \frac{1}{(x+1)^2}$$

3) $\dfrac{4x^2 + x + 4}{(x-1)(x^2+x+1)}$

Solution. Form the sum:

$$\frac{A}{x-1} + \frac{Bx+C}{x^2+x+1} = \frac{A(x^2+x+1) + (Bx+C)(x-1)}{(x-1)(x^2+x+1)}$$

Set the resulting numerator equal to the given numerator:

$$4x^2 + x + 4 = (A+B)x^2 + (A+C-B)x + A - C$$

Then we have

$$\begin{array}{c|c} A + B = 4 & A = 3 \\ A - B + C = 1 & B = 1 \\ A - C = 4 & C = -1 \end{array}$$

$$\therefore \quad \frac{4x^2 + x + 4}{(x-1)(x^2+x+1)} = \frac{3}{x-1} + \frac{x-1}{x^2+x+1}$$

4) $\dfrac{x^3 - 2x^2 + 3x}{(x^2+2x+3)^2}$

Solution. Form the sum:

$$\frac{Ax+B}{x^2+2x+3} + \frac{Cx+D}{(x^2+2x+3)^2}$$

The numerator of this sum is

$$(Ax+B)(x^2+2x+3) + Cx + D$$
$$= Ax^3 + (2A+B)x^2 + (3A+2B+C)x + 3B + D$$

and we can write

$$A = 1$$
$$2A + B = -2$$
$$3A + 2B + C = 3$$
$$3B + D = 0$$

$$A = 1$$
$$B = -4$$
$$C = 8$$
$$D = 12$$

$$\therefore \quad \frac{x^3 - 2x^2 + 3x}{(x^2 + 2x + 3)} = \frac{x - 4}{x^2 + 2x + 3} + \frac{8x + 12}{(x^2 + 2x + 3)^2}$$

Summary of Symbols

$P(x) = a_0 x^n + a_1 x^{n-1} + \ldots + a_{n-1} x + a_n$, polynominal of degree n

$a + \sqrt{b}$, quadratic surd

$D_x y$, derivative of y with respect to x

$\displaystyle\sum_{i=1}^{n}$, the summation from $i = 1$ to n

---

74. Exercise VIII.

1. Prove the Remainder Theorem.
2. Determine a_1 and a_2 so that
   a) $x^4 + a_1 x^3 + a_2 x^2 + 9x - 36$ is divisible by $(x + 3)(x - 4)$.
   b) $x^4 + a_1 x^3 + a_2 x^2 - 7x - 6$ is divisible by $(x + 2)(x - 3)$.
   c) $x^3 + a_1 x^2 + a_2 x + 2$ is divisible by $(x + 1)(x - 2)$.
   d) $6x^4 + a_1 x^3 - 8x^2 + a_2 x + 6$ is divisible by $(2x - 3)(x + 1)$.
   e) $2x^4 + a_1 x^3 + a_2 x^2 + 26x + 2$ is divisible by $(2x - 4)(x - 1)$.
3. Find the solution set of each equation:
   a) $x^3 - 7x + 6 = 0$
   b) $4x^3 - 8x^2 + 5x - 1 = 0$
   c) $x^3 - 4x^2 - 17x + 60 = 0$
   d) $4x^3 + 2x^2 - 4x + 1 = 0$
   e) $6x^3 - 7x^2 - 3x = 0$
   f) $x^4 + 3x^3 - x - 3 = 0$
   g) $x^4 + 4x^3 + 3x^2 + 4x + 12 = 0$
   h) $x^3 - 3x^2 - 6x + 8 = 0$
   i) $x^4 - 6x^3 + 13x^2 - 14x + 6 = 0$
   j) $x^4 - 5x^3 + 20x - 16 = 0$
   k) $2x^4 + 3x^3 - 6x^2 + 8x - 3 = 0$
   l) $x^3 + x^2 + x + 1 = 0$
   m) $x^4 - x^3 - 27x^2 + 25x + 50 = 0$
   n) $4x^4 + 8x^3 - 7x^2 - 21x - 9 = 0$
   o) $16x^5 + 12x^4 - 12x^3 - 37x^2 - 33x - 9 = 0$
   p) $x^3 = 1$

q) $12x^6 - 44x^5 + 41x^4 - 5x^3 + 13x^2 - 11x - 6 = 0$

4. Find the relative extrema of the following functions:

    a) $x^3 - 2x^2 + x - 5$ b) $x^3 - 4x - 4$

    c) $2x^3 - 2x^2 - 2x + 3$ d) $x^2 - 2x + 4$

    e) $9x^4 - 10x^3 - 3x^2 + 6x + 1$

5. Sketch the graphs of the following polynomials:

    a) The functions in Problem 4   b) $y = x^4 - x^2$

    c) $y = x^3 + 5x^2 + 3x - 6$ d) $y = x^2y + 2x$

    e) $yx = x + 3$

6. Solve the following inequalities:

    a) $x^2 + x - 6 > 0$ b) $x^2 + 2x + 1 \geq 4$

    c) $(x - 3)(x + 2)(x - 1) \leq 0$ d) $x^2 + 9 < 25$

    e) $(x - 2)^2(x + 1) \geq 0$ f) $(x + 4)^2 < 9$

7. Draw the graphs and shade the set of points satisfying the following inequalities:

    a) $x^2 + y^2 < 4$ and $x + y > 1$

    b) $(x - 2)^2 + (y - 2)^2 < 4$ and $x + y < 3$

    c) $y - x^2 + x < 6$ and $y < 3$

    d) $y^2 < 4x, \; x < 4,$ and $y < 2$

    e) $x^2 + y^2 < 5$ and $y^2 - 4x < 0$

8. Resolve the following fractions into their simplest partial fractions:

    a) $\dfrac{2x^2 - 17x - 24}{4x^3 + 4x^2 - 9x - 9}$

    b) $\dfrac{3x^4 + 6x^3 - 2x^2 + 5x + 6}{x^3 + 2x^2 - x - 2}$

    c) $\dfrac{2x^2 - 2x + 1}{x^4 - 2x^3 + x^2}$

    d) $\dfrac{2x^2 - 5x + 7}{x^3 - 6x^2 + 11x - 6}$

    e) $\dfrac{x^2 + 1}{x^4 - x^3}$

    f) $\dfrac{x^2 + 1}{x^2 - 2x + 1}$

    g) $\dfrac{3x - 4}{(x^2 - 1)(x^4 + 2x^2 + 1)}$

    h) $\dfrac{3x^2 - 4x + 4}{x^3 - 1}$

    i) $\dfrac{x^4 + x^3 + 2x^2 - 7}{(x + 2)(x^2 + x + 1)^2}$

j) $\dfrac{x^3 + 2x^2 - 1}{x^4 + 2x^2 + 1}$

k) $\dfrac{1}{x^3 + x}$

l) $\dfrac{3x + 4}{x^4 + x^3 + x^2}$

9. Use the graphical method to approximate the three roots to the nearest tenth:

$$8x^3 - 12x^2 + 1 = 0$$

10. Plot the function defined by

$$y = \begin{cases} [-3, 3] & \text{when} \quad x = 0 \\ -\dfrac{3}{2}x + 3 & \text{when} \quad 0 < x < 4 \\[2mm] [-3, 3] & \text{when} \quad x = 4 \\ \pm\sqrt{9 - (x - 9)^2} & \text{when} \quad 6 \le x \le 12 \end{cases}$$

11. Use the same set of axes and sketch the functions defined by the following:

a) $y = \dfrac{1}{2}(x + 7)$ $\qquad\qquad$ $d_f : [-2, -1]$

b) $y = -\dfrac{1}{2}(x - 7)$ $\qquad\qquad$ $d_f : [1, 2]$

c) $y = -\dfrac{3}{2}(x - 5)$ $\qquad\qquad$ $d_f : [3, 5]$

d) $y = \dfrac{3}{2}(x + 5)$ $\qquad\qquad$ $d_f : [-5, -3]$

e) $y = -2 + |x|$ $\qquad\qquad$ $d_f : [-1, 1]$

f) $y = \dfrac{1}{3}\sqrt{225 - 16x^2}$ $\qquad\qquad$ $d_f : [-3, 3]$

g) $y = -\dfrac{1}{5}\sqrt{400 - 16x^2}$ $\qquad\qquad$ $d_f : [-5, 5]$

12. Plot the function defined by

$$y = \dfrac{8}{4 + x^2}$$

13. Plot the functions defined by the following:
   a) $y = x^n$ for $n = 1, 2, 3, 4,$ and 5
   b) $y = x^n$ for $n = 1, 1/2,$ and $1/3$

VII

EXPONENTIAL AND LOGARITHMIC FUNCTIONS

75. Introduction. In the last chapter we discussed functions defined by polynomials. The particular function defined by

$$y = x^n$$

where x is the independent variable and n is a rational number is called a **power function.** Suppose we interchange the role of the variables and constants and consider the function defined by

$$y = a^x$$

where a is a constant and x is still the argument. In this chapter we shall discuss functions of this type and their inverses.

76. The Exponential Function. We shall limit our discussion to the real-number system, which requires that the values of the parameter a be positive; i.e., $a > 0$.

Definition. *The function* f *defined by the equation* $y = a^x$, $(a > 0)$, $x \in R$, *is called the* **exponential function** *with base* a.

Since the domain of definition is the set of real numbers, the range is $0 < y < \infty$. Let us demonstrate this with a few examples. Let $a = 5, 3, 2, 1$, and $1/2$, and calculate the tables of values (p. 156). The graphs* are shown in Figure 46 (p. 157).

Let us study these graphs and notice some of their characteristics.
- a) The function is positive, $y > 0$, for all values of x.
- b) When $x = 0$, the function has the value 1, $y = 1$, for all $a > 0$.

---

*In drawing the graphs we join the points by a smooth curve, which assumes that for irrational values of x the values of y will be such that the corresponding points (x, y) will lie on the curve. The existence of values for y can be shown by the use of the least upper bound principle of a sequence of numbers. Such detailed analysis can be found in more advanced books.

x	$y = 5^x$	$y = 3^x$	$y = 2^x$	$y = 1^x$	$y = \left(\dfrac{1}{2}\right)^x$
-3	$\dfrac{1}{125}$	$\dfrac{1}{27}$	$\dfrac{1}{8}$	1	8
-2	$\dfrac{1}{25}$	$\dfrac{1}{9}$	$\dfrac{1}{4}$	1	4
-1	$\dfrac{1}{5}$	$\dfrac{1}{3}$	$\dfrac{1}{2}$	1	2
0	1	1	1	1	1
1	5	3	2	1	$\dfrac{1}{2}$
2	25	9	4	1	$\dfrac{1}{4}$
3	125	27	8	1	$\dfrac{1}{8}$

c) Since $y > 0$, there is no zero of the function.

d) The X-axis, $y = 0$, is an asymptote.

e) If $a = 1$, then y is a constant, $y = 1$.

f) If $a > 1$, the function is an increasing function; i.e., as x increases so does y; such functions are called *monotone increasing*.

g) If $a < 1$, the function decreases as x increases; this is a *monotone decreasing* function.

Numbers which cannot be obtained as roots of algebraic equations with rational coefficients are called *transcendental numbers*. The well known π is such a number. We shall consider another one defined by

$$e = \lim_{n \to 0} (1 + n)^{\frac{1}{n}}$$

Although it requires a knowledge of calculus to prove that this number exists, we shall calculate some values around $n = 0$ in Section 78 and show that e lies between certain values. Like the number π, its value can be calculated to as many digits as desired.

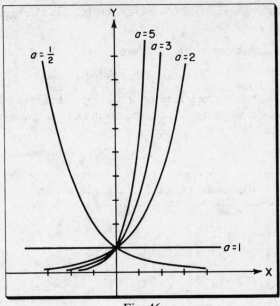

Fig. 46

The first eleven are

$$e = 2.71828\ 18285$$

If we let $a = e$, the exponential function takes the form

$$y = e^x$$

This form is so important in the application of mathematics that it is usually referred to as *the* exponential function. Values of this function have been calculated and extensive tables have been published.* A very limited table is given in the back of this book. The graph of the function is shown in Figure 48 (p. 159).

77. The Logarithmic Function.   The exponential function $y = a^x$ is a strictly monotone function for all positive values of a except $a = 1$. Consequently, if we restrict the values of a to $a > 0$, $a \neq 1$, we have a unique value of y for each value of x and the inverse function exists. To find an expression for the inverse function we need to solve the equation $x = a^y$ for y. This necessitates some new terminology.

Definition.   *The inverse of the function* $y = a^x$, $a > 0$, $a \neq 1$,

---

*For a medium table see Kaj L. Nielsen, *Logarithmic and Trigonometric Tables,* Revised Edition, New York: Barnes and Noble, Inc., 1962.

is given by the expression $y = \log_a x$ *and is called the* **logarithm** *of* x *to the base* a.

With this definition we can then state

$$\text{If } x = a^y, \text{ then } y = \log_a x$$

Under the restrictions placed on a, the logarithm is unique. The domain is the set of positive real numbers and the range is the set of all real numbers.

Examples:

Exponential Form	Logarithmic Form
$3^2 = 9$	$\log_3 9 = 2$
$2^3 = 8$	$\log_2 8 = 3$
$3^{-2} = \dfrac{1}{9}$	$\log_3 \dfrac{1}{9} = -2$
$16^{\frac{1}{2}} = 4$	$\log_{16} 4 = \dfrac{1}{2}$

The graph of $y = \log_a x$ can be obtained for a specified value of a by plotting pairs (x, y) or by reflecting the graph of $y = a^x$ in the line $y = x$. Values for (x, y) can be found from logarithmic tables or can be calculated by giving values to y and finding x from the equation $x = a^y$. The graphs for $a = 3$ and $a = 1/2$ are shown in Figure 47.

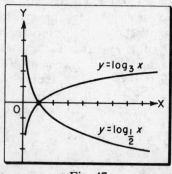

Fig. 47

Since there is a one-to-one correspondence between the positive numbers and their logarithms, it follows that if two numbers are equal, then their logarithms are equal. Thus, if $M = N > 0$, then

$$\log_a M = \log_a N$$

The two most important values of a are $a = 10$ and $a = e$. When $a = 10$, the logarithms are called *common logarithms* and we omit the base in writing $\log x$. If $a = e$, the logarithms are called *natural logarithms* and we write $\ln x$ for the symbolic notation. Extensive tables for the logarithms of numbers to these two bases have been published. Limited tables are given in the back of this book. The graphs of $y = e^x$ and $y = \ln x$ are shown in Figure 48.

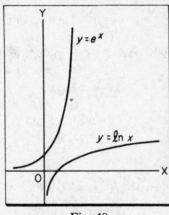

Fig. 48

78. Properties of Logarithms.

We shall use the laws of exponents given in Section 6 to develop three important properties of logarithms.

Property 1. *The logarithm of a product is equal to the sum of the logarithms of its factors.*

$$\log_a MN = \log_a M + \log_a N$$

Proof.

Let $\qquad m = \log_a M \qquad$ and $\qquad n = \log_a N$

By definition: $\qquad M = a^m \qquad$ and $\qquad N = a^n$

Then $\qquad\qquad MN = a^m a^n = a^{m+n}$

By definition: $\log_a MN = m + n = \log_a M + \log_a N$

Property 2. *The logarithm of a quotient is equal to the logarithm of the numerator minus the logarithm of the denominator.*

$$\log_a \frac{M}{N} = \log_a M - \log_a N$$

Proof.

Let $\qquad\qquad m = \log_a M \qquad$ and $\qquad n = \log_a N$

By definition: $\qquad M = a^m \qquad$ and $\qquad N = a^n$

Then $\qquad\qquad\qquad \dfrac{M}{N} = \dfrac{a^m}{a^n} = a^{m-n}$

By definition: $\quad \log_a \dfrac{M}{N} = m - n = \log_a M - \log_a N$

Property 3.   *The logarithm of the* k*th power of a number equals* k *times the logarithm of the number.*

$$\log_a N^k = k \log_a N$$

Proof.

Let $\qquad\qquad\qquad n = \log_a N$

By definition: $\qquad\qquad N = a^n$

By the laws of exponents: $\quad N^k = (a^n)^k = a^{kn}$

By definition: $\qquad \log_a N^k = kn = k \log_a N$

These properties are used extensively in computation and in the solution of equations.

Illustration.   Express

$$\log_a \frac{x^2 \sqrt{y}}{z^3 w}$$

as the algebraic sum of logarithms.

Solution.   By Property 2:

$$\log_a \frac{x^2 \sqrt{y}}{z^3 w} = \log_a(x^2 \sqrt{y}) - \log_a(z^3 w)$$

By Property 1: $\qquad = \log_a x^2 + \log_a \sqrt{y} - (\log_a z^3 + \log_a w)$

By Property 3: $\qquad = 2 \log_a x + \dfrac{1}{2} \log_a y - 3 \log_a z - \log_a w$

Let us consider the number

$$N = (1 + n)^{\frac{1}{n}}$$

for small values of n. Take the logarithm of both sides and apply Property 3, to obtain

$$\log N = \frac{1}{n} \log (1 + n)$$

We can now calculate the following table:

n	$- .5$	$- .2$	$- .1$	$- .01$	$.01$	$.1$	$.2$	1
N	4.00	3.05	2.87	2.729	2.704	2.59	2.49	2.00

It appears that the value at $n = 0$ should be $2.704 < N_1 < 2.729$. If our logarithmic tables were sufficiently large, we could continue the calculations near $n = 0$ and obtain more accurate approximations of the value of e.

The following characteristics are direct consequences of the definition.

a) $\log_a a = 1$
b) $a^{\log_a x} = x$
c) $\log_a a^x = x$

79. Change of Base. The logarithmic function is defined in terms of the base as a parameter. There is a very simple relationship between logarithms to different bases.

Property.   *The logarithm of a number* N *to the base* b *is equal to the quotient of its logarithm to the base* a *divided by the logarithm of* b *to the base* a.

$$\log_b N = \frac{\log_a N}{\log_a b}$$

Proof.   Let $x = \log_b N$; then by definition $N = b^x$. Take the logarithm of both members to the base a:

$$\log_a N = \log_a b^x = x \log_a b$$

Solve for x:
$$x = \frac{\log_a N}{\log_a b}$$

Replace x by $\log_b N$: $\log_b N = \dfrac{\log_a N}{\log_a b}$

Let us apply this property to the bases 10 and e.

$$\log N = \frac{\ln N}{\ln 10} \quad \text{and} \quad \ln N = \frac{\log N}{\log e}$$

The values of the constants to four decimals are

$$\log e = 0.4343 \quad \text{and} \quad \ln 10 = 2.3026$$

which are reciprocals of each other. Thus

$$\log N = 0.4343 \ln N \quad \text{and} \quad \ln N = 2.3026 \log N$$

Illustrations.
1) Find $\log_8 6$.

Solution. $\log_8 6 = \dfrac{\log 6}{\log 8} = \dfrac{0.7782}{0.9031} = 0.8617$

2) Use base 10 to find $\ln 12$ and check by table.
Solution. $\ln 12 = 2.3026 \log 12 = 2.3026(1.0792)$
        $= 2.4850$

3) Find $\log_2 4$.

Solution. $\log_2 4 = \dfrac{\log 4}{\log 2} = \dfrac{0.6021}{0.3010} = 2.000$

80. Exponential and Logarithmic Equations. We are now in a position to consider two types of transcendental equations.

Definition.   *If a variable in an equation occurs as an exponent, the equation is called an* **exponential equation.** *If a variable in an equation occurs in an expression whose logarithm is indicated, the equation is called a* **logarithmic equation.**

Examples:
a) $3^x = 27$ and $2^x = 3^{x+1}$ are exponential equations in one variable.
b) $y = 1/2(e^x - e^{-x})$ is an exponential equation in two variables.
c) $\log(x^2 + 5x) = 3$ is a logarithmic equation in one variable.
d) $y = \ln(x + \sqrt{x+1})$ is a logarithmic equation in two variables.

We are usually concerned with finding the solution sets of or analyzing the functions defined by these equations. The exponential equations may be solved by reducing the equation to a common base or by taking the logarithm of both sides.

Illustrations.

1) Solve $3^{x+2} = 81$.

Solution.   Since $81 = 3^4$ we have

$$3^{x+2} = 81 = 3^4$$

and consequently

$$x + 2 = 4 \quad \text{or} \quad x = 2$$

2) Solve $(2^{x+1})(3^{2x}) = 6^{2x+1}$.

Solution.   Take the logarithm of both sides to obtain

$$\log [(2^{x+1})(3^{2x})] = \log (6^{2x+1})$$

Apply the properties of logarithms and collect terms:

$$(x + 1) \log 2 + 2x \log 3 = (2x + 1) \log 6$$
$$(\log 2 + 2 \log 3 - 2 \log 6)x = \log 6 - \log 2$$
$$[\log \frac{(2)(3^2)}{36}]x = \log \frac{6}{2}$$
$$x = \frac{\log 3}{\log \frac{1}{2}} = -\frac{\log 3}{\log 2}$$
$$= -\frac{0.4771}{0.3010} \doteq -1.59$$

3) Discuss the function defined by

$$y = \frac{e^x + e^{-x}}{2}$$

and sketch its graph.

Solution.   Since $e^x > 0$ and $e^{-x} = \frac{1}{e^x} > 0$ for all real values of x, then (see Section 72)

$$e^x + e^{-x} \geq 2 \quad \text{and} \quad y \geq 1$$

The graph is symmetric with respect to the Y-axis since the equation is the same if x is replaced by $-x$. Some typical values are (use the table in back of the book)

x	0	$\pm .5$	± 1	± 1.5	± 2	± 2.5	± 3
y	1	1.13	1.54	2.35	3.76	6.13	10.07

The curve, which is called a *catenary*, is shown in Figure 49 (p. 164).

Logarithmic equations may be solved by applying the properties of logarithms and the definition.

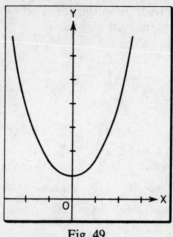

Fig. 49

Illustrations.

1)   Solve: $\log x + \log (x + 48) = 2$.
Solution.

By Property 1: $\log [x(x + 48)] = 2$
By definition: $x^2 + 48x = 10^2 = 100$
Solving for x: $x^2 + 48x - 100 = 0$
           $(x - 2)(x + 50) = 0$

$$x = 2 \quad \text{and} \quad x = -50$$

Since logarithms are not defined for negative numbers, $x = -50$ is not a solution.

Check. $\log 2 + \log 50 = 2$

$$0.3010 + 1.6990 = 2$$

2) Solve $\ln x + \ln (3x - 6e + 2) = 1 + \ln 4$
Solution.   Perform the following steps by using the properties and definition:

$$\ln x + \ln (3x - 6e + 2) - \ln 4 = 1$$

$$\ln \left[\frac{3x^2 + (2 - 6e)x}{4} \right] = 1$$

$$3x^2 + (2 - 6e)x = 4e$$

$$3x^2 + (2 - 6e)x - 4e = 0$$

$$(3x + 2)(x - 2e) = 0$$

The solution is $x = 2e$.

3) Find the inverse of the function defined by

$$y = \ln x + \ln(x + 2)$$

Solution. The equation may be written

$$y = \ln x(x + 2)$$

and to find the inverse we need to solve

$$x = \ln y(y + 2)$$

for y in terms of x. By definition

$$e^x = y^2 + 2y$$

so that

$$y = \frac{-2 \pm \sqrt{4 + 4e^x}}{2} = -1 \pm \sqrt{1 + e^x}$$

The domain of the given function is $x > 0$; therefore the range of the inverse is $y > 0$. Since $e^x > 0$ for all x, we have $1 + e^x > 1$ for all x. Thus $y = -1 + \sqrt{1 + e^x} > 0$ for all x and defines the inverse f^{-1}.

4) Transform the equation

$$\ln y - 3x = 2 \ln x$$

into an equation free of logarithms.
Solution. Collect terms to obtain

$$\ln y - 2 \ln x = 3x$$

$$\ln y - \ln x^2 = 3x$$

$$\ln \frac{y}{x^2} = 3x$$

Then by definition

$$\frac{y}{x^2} = e^{3x} \qquad \text{or} \qquad y = x^2 e^{3x}$$

81. Hyperbolic Functions. In the last section we used the function defined by

$$y = \frac{e^x + e^{-x}}{2}$$

in one of the illustrations. This function is one of six which form a special class called the *hyperbolic functions*. The first three are defined by

$$\sinh x = \frac{e^x - e^{-x}}{2},$$

$$\cosh x = \frac{e^x + e^{-x}}{2}$$

$$\tanh x = \frac{e^x - e^{-x}}{e^x + e^{-x}}$$

The other three are defined as the reciprocals of these. Let us consider the following abbreviated table of values.

x	-2.5	-2	-1.5	-1	0	1	1.5	2.0	2.5
e^x	0.08	0.14	0.22	0.37	1	2.72	4.48	7.39	12.18
e^{-x}	12.18	7.39	4.48	2.72	1	0.37	0.22	0.14	0.08
$e^x - e^{-x}$	-12.10	-7.25	-4.26	-2.35	0	2.35	4.26	7.25	12.10
$e^x + e^{-x}$	12.25	7.53	4.70	3.09	2	3.09	4.70	7.53	12.26
$\sinh x$	-6.05	-3.63	-2.13	-1.18	0	1.18	2.13	3.63	6.05
$\cosh x$	6.13	3.76	2.35	1.54	1	1.54	2.35	3.76	6.13
$\tanh x$	-0.99	-0.96	-0.91	-0.76	0	0.76	0.91	0.96	0.99

These three functions are defined for all x and
  $\sinh x$ has the range $]-\infty, \infty[$
  $\cosh x$ has the range $[1, \infty[$
  $\tanh x$ has the range $]-1, 1[$
The inverse of the function defined by

$$y = \sinh x = \frac{1}{2}(e^x - e^{-x})$$

is denoted by $\sinh^{-1} x$ and is found by solving the equation

$$x = \frac{1}{2}(e^y - e^{-y})$$

for y. The equation can be simplified into the form

$$e^{2y} - 2xe^y - 1 = 0$$

with solution

$$e^y = x \pm \sqrt{1 + x^2}$$

which in turn must be positive for all x. Since $\sqrt{1 + x^2} > x$ for all x, the solution $x - \sqrt{1 + x^2} < 0$ for all x is an extraneous solution. We therefore have

$$y = \sinh^{-1} x = \ln(x + \sqrt{1 + x^2})$$

82. Logarithmic Graph Paper. A logarithmic scale can be constructed by dividing a given length according to the logarithm of the

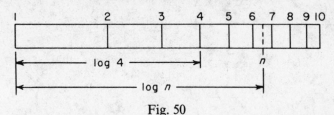

Fig. 50

scaled number. See Figure 50. Two logarithmic scales placed on rulers may be combined to perform multiplication and division; this is the basis for the slide rule.*

If we scale both axes of a rectangular coordinate system logarithmically, we form a *log-log graph paper*. If we scale one axis logarithmically and the other axis linearly we form a *semilog graph paper*. Both types are used to analyze experimental data.

Consider the three functions defined by

  a) $y = ax + b$, a linear function,
  b) $y = ax^n$, a power function,
  c) $y = ae^{bx}$, an exponential function.

We have already studied the linear function and shown that its graph is a straight line. Apply the properties of logarithms to the other two functions.

$$y = ax^n \qquad\qquad\qquad y = ae^{bx}$$
$$\log y = \log a + n \log x \qquad \log y = \log a + bx \log e$$
$$Y = A + nX \qquad\qquad\qquad Y = A + Bx$$

where $Y = \log y$, $X = \log x$, $A = \log a$, and $B = b \log e$. Since a, b, and e are constants, their logarithms are constants and A and B are constants. The equations in terms of the new variables X and Y are therefore linear equations and their graphs will be straight lines in their respective coordinate systems. They may be plotted directly from the original equations using the appropriate graph paper. The graph of the equation $y = ax^n$ will be a straight line when plotted on log-log paper. The graph of the equation $y = ae^{bx}$ will be a straight line when plotted on semilog paper.

Illustrations.

1) Obtain a straight-line graph of the equation $y = 2x^2$.
Solution. Calculate the table of values.

---

*See Calvin Bishop, *Slide Rule*, New York: Barnes and Noble, Inc., 1958.

x	1	2	3	4	5	10
y	2	8	18	32	50	200

Plot these values on log-log paper. The graph is shown in Figure 51.

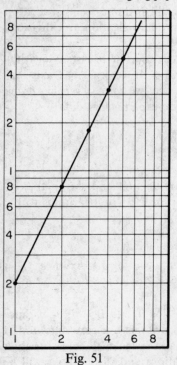

Fig. 51

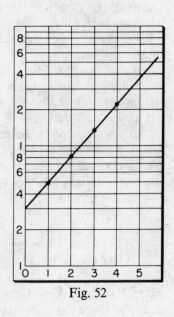

Fig. 52

2) Obtain a straight-line graph of the equation $y = 3e^{\frac{x}{2}}$.

Solution.   Use the exponential table to calculate the values.

x	0	1	2	3	4
y	3	4.9	8.2	13.4	22.2

Plot these values on semilog paper. See Figure 52.

Summary of Symbols	
$e \doteq 2.71828 +$, the exponential number $\log_a N$, logaritum of N to the base a $\log N$, logarithm of N to the base 10 $\ln N$, logarithm of N to the base e	$\exp(x) = e^x$ $\sinh x$, hyperbolic sine of x $\cosh x$, hyperbolic cosine of x $\tanh x$, hyperbolic tangent of x

83. Exercise IX.

1. Sketch the graphs of the following equations:

   a) $y = (2)(3^x)$ b) $y = 2e^x$

   c) $y = \dfrac{1}{2}e^{2x}$ d) $y = \dfrac{1}{2}e^{-2x}$

   e) $y = \dfrac{1}{2}(e^x - e^{-x})$ f) $y = \log_2 x$

   g) $y = \ln x$ h) $y = \log x$

2. Use the properties of logarithms to express the logarithm of each of the following numbers as the algebraic sum of logarithms:

   a) $N = x^3\sqrt{y}$ b) $N = 3\sqrt{2.9}$

   c) $N = \dfrac{32.3\sqrt{5.18}}{(12)(4.9)^2}$ d) $N = \dfrac{gV^2}{t^3\sqrt{h}}$

3. Find the following values:

   a) $\log_2 7$ b) $\log_3 8$

   c) $\log_2 8$ d) $\log_9 2$

   e) $\log_5 9$ f) $\log_{16} 4$

4. Solve the following equations:

   a) $4^x = 17$ b) $4^x = \dfrac{1}{64}$

   c) $(3^{2x})(2^x) = 16$ d) $8^{x^2} = 64$

   e) $2^{x^2 - 3x} = \dfrac{1}{2}$ f) $e^{2x} = 4$

   g) $\log(x - 5) + \log(3x) = 1$

   h) $\log(x + 3) + \log(x - 2) = 2\log 6$

   i) $\log(25x + 3) - \log 2x = 2$

5. What is the domain of definition for each of the following? (Hint: Since $N > 0$ for $\log N$ to exist, both numerator and denominator must be simultaneously positive or negative if

N is a fraction.)

a) $\log \dfrac{2x - 1}{x}$

b) $\log \dfrac{3x + 6}{5x}$

c) $\log (3x - 4)$

d) $\log \dfrac{x}{3x - 6}$

6. What is the domain for which the following logarithms are positive? (Hint: $\log N > 0$ if $N > 1$.)

    a) $\log (3x - 5)$ b) $\log (2x + 1)$

    c) $\log \left(\dfrac{2x + 1}{3x} \right)$ d) $\log \left(\dfrac{2x}{3x - 6} \right)$

7. Use the definitions to prove the following identities:

    a) $\sinh (x + y) = \sinh x \cosh y + \cosh x \sinh y$

    b) $\sinh 2x = 2 \sinh x \cosh x$

    c) $\cosh^2 x - \sinh^2 x = 1$

    d) $\cosh x + \sinh x = e^x$

    e) $\sinh (- x) = - \sinh x$

8. Find the inverse of $y = \cosh x$.

9. Obtain straight-line graphs of the following equations:

    a) $y = 3x^2$ b) $y = 2x^3$

    c) $y = 2e^{2x}$ d) $y = \dfrac{1}{3} e^{3x}$

    e) $y = 3x - 5$ f) $y = 2x^{\frac{1}{2}}$

10. Sketch the graph of $y = x^x$.

VIII

SERIES

84. Definitions.   If in a set of n objects we select one object to be the first, another to be the second, still another the third, etc., we have an ordered set which may be represented by

$$a_1, a_2, a_3, \ldots, a_n, \ldots$$

An ordered set of objects is called a **sequence** and the individual members are called **terms** or **elements** of the sequence. A common representation of a finite sequence, one that has a finite number of terms, is

$$a_i, (i = 1, 2, 3, \ldots, n)$$

which is read "a_1 for i equal 1 to n." If the sequence has an infinite number of terms,* we may write

$$a_i, (i = 1, 2, 3, \ldots)$$

If the general term, a_i, has some law of formation, we can then define the sequence by the general term. Since a sequence is an ordered set, we can also describe it in set notation; e.g.,

$$A = \{a_i | a_i = 2i - 1, (i = 1, 2, 3, \ldots, 9)\}$$
$$= 1, 3, 5, 7, 9, 11, 13, 15, 17$$

The reader may already be familiar with the following two sequences of numbers.

Definition.   *An* **arithmetic progression (A.P.)** *is a sequence of numbers in which each term, after the first, is obtained from the preceding one by adding to that term a fixed number called the* **common difference.**

Let a denote the first term,
   d denote the common difference,

---

*If after each term of a sequence there exists another term, it is called an **infinite sequence**. See Section 33.

171

n denote the number of terms,

b_i denote the ith term,

l denote the nth (last) term.

Then the A.P. is $a, a + d, a + 2d, \ldots, a + (i - 1)d, \ldots, l$.

Note that

$$b_i = a + (i - 1)d \quad \text{and} \quad l = a + (n - 1)d$$

and that the A.P may be extended in either direction by simply adding or subtracting d.

The terms between two given terms of an A.P. are called **arithmetic means**.

Illustrations.

1) Write the first seven terms of the A.P. if $a = 2$ and $d = 3$.

Solution.   2, 5, 8, 11, 14, 17, 20

Check. $l = 2 + 6(3) = 2 + 18 = 20$

2) Insert five arithmetic means between 5 and 8.

Solution. $a = 5, l = 8, n = 2 + 5 = 7$

Then since $l = a + (n - 1)d$

we have $8 = 5 + 6d \quad \text{or} \quad d = \dfrac{1}{2}$

The A.P. is

$$5, \frac{11}{2}, 6, \frac{13}{2}, 7, \frac{15}{2}, 8$$

3) Extend the A.P. 1, 5, 9 two terms in each direction.

Solution.   Since the second term is $a + d$, we have

$$5 = 1 + d \quad \text{or} \quad d = 4$$

The new A.P. is

$$-7, -3, 1, 5, 9, 13, 17$$

Definition.   *A* **geometric progression (G.P.)** *is a sequence of numbers in which each term, after the first, is obtained from the preceding one by multiplying that term by a fixed number called the* **common ratio.**

Let us use the same notation as we did for the A.P., except let r denote the common ratio. Then the G.P. is

$$a, ar, ar^2, \ldots, ar^{i-1}, \ldots, l = ar^{n-1}$$

The terms between two terms of a G.P. are called **geometric means**.

Illustrations.

1) Find the seventh term of the G.P. which has $a = 3$ and $r = 2$.

Solution.   The seventh term is given by ar^6; therefore we have

$$ar^6 = 3(2)^6 = 3(64) = 192$$

2) Given $a = -2, n = 7$, and $l = -1/32$. Find r.

Solution.   Since $l = ar^{n-1}$, we have

$$-\frac{1}{32} = (-2)r^6$$

$$\frac{1}{64} = r^6$$

$$\pm\frac{1}{2} = r$$

There are two geometric progressions:

$$-2, -1, -\frac{1}{2}, -\frac{1}{4}, -\frac{1}{8}, -\frac{1}{16}, -\frac{1}{32}$$

$$-2, 1, -\frac{1}{2}, \frac{1}{4}, -\frac{1}{8}, \frac{1}{16}, -\frac{1}{32}$$

3. Insert four geometric means between 3 and 96.

Solution. $a = 3, l = 96, n = 2 + 4 = 6$

Then

$$l = ar^{n-1}$$
$$96 = 3r^5$$
$$32 = r^5$$
$$2 = r$$

The G.P. is $3, 6, 12, 24, 48, 96$

The infinite geometric progression is an interesting sequence. Consider the G.P. defined by

$$b_i = \frac{1}{2^i}, \quad (i = 1, 2, 3, \ldots)$$

the first few terms of which are

$$\frac{1}{2}, \frac{1}{4}, \frac{1}{8}, \frac{1}{16}, \frac{1}{32}, \frac{1}{64}, \cdots$$

We notice that the numbers are getting smaller but are always positive. They are therefore approaching the number 0 from the positive side. This is an illustration of the concept of a limit.

Definition.   *A sequence* a_i, $(i = 1, 2, 3, \ldots)$, *is said to approach the constant* c *as a* **limit** *if the value of* $|c - a_i|$ *becomes and remains less than any preassigned positive number, however small.*

This is expressed symbolically by

$$\lim_{n \to \infty} a_n = c$$

which is read "the limit of a_n as n increases without bound is c."

If a sequence approaches a limit, it is said to be **convergent**; if it does *not* approach a limit, it is said to be **divergent**.

If the nth term of a sequence approaches 0, then the limit exists and the sequence is convergent. In particular if $|r| < 1$ in a geometric progression, the sequence converges. Any geometric sequence for which $|r| > 1$ forms an example of a divergent sequence.

The concept of limits is very important in mathematics. Let us consider it a little further. We say that the general term of the infinite sequence defined by

$$a_n = \frac{1}{n}, \quad (n = 1, 2, 3, \ldots)$$

approaches 0 as a limit. If this statement is true, the above definition specifies that if we choose a small positive number, $\varepsilon > 0$, (say $\varepsilon = 10^{-10}$) then there is a value of n, say n_0, such that for all n larger than n_0, we have $1/n < \varepsilon$. Let us consider a formal proof.

Theorem. *For every ε, there exists an* n_0 *such that* $1/n < \varepsilon$ *for all* $n > n_0$. (This is equivalent to $\lim_{n \to \infty} (1/n) = 0$.)

Proof. Choose the small positive number ε. We want $1/n < \varepsilon$ or (by the properties of inequalities) $n > 1/\varepsilon$. Now choose $n_0 = 1/\varepsilon$. It follows (again by the properties of inequalities) that if $n > n_0$, then $n > 1/\varepsilon$ and hence $1/n < \varepsilon$. In case $1/\varepsilon$ is not an integer, we choose n_0 to be the first integer just larger than $1/\varepsilon$, which again assures that $n > n_0 > 1/\varepsilon$.

We now state a more formal definition of convergence.

Definition. *If for every positive number ε, there exists a real number* n_0 *such that*

$$|S_n - S| < \varepsilon \quad \text{for } n > n_0$$

then the sequence S_n *is said to* **converge** *to the limit* S.

85. Series. We shall now consider the sum of the terms of a sequence.

Definition. *The sum of the terms of a sequence is called a* **series**.

We shall use two symbols to denote the sum:

$$S_n = a_1 + a_2 + a_3 + \ldots + a_n = \sum_{i=1}^{n} a_i$$

If the sequence is finite the sum will always exist.

Illustrations.
1) Find the sum of the first n terms of an A.P.
Solution.

$$\begin{aligned} S_n &= a + (a + d) + (a + 2d) + \ldots + [a + (n - 1)d] \\ &= (a + a + \ldots + a) + [d + 2d + 3d + \ldots (n - 1)d] \\ &= na + d[1 + 2 + 3 + \ldots + (n - 1)] \end{aligned}$$

The sum in the brackets (see Illustration, page 61) is $\frac{1}{2}(n - 1)n$ so we have

$$\begin{aligned} S_n &= na + \frac{1}{2}n(n - 1)d \\ &= \frac{n}{2}[2a + (n - 1)d] \\ &= \frac{n}{2}(a + l) \end{aligned}$$

2) Find the sum of the first n terms of a G.P.
Solution. Write the following two equations:

$$S_n = a + ar + ar^2 + \ldots + ar^{n-1}$$
$$rS_n = ar + ar^2 + \ldots + ar^{n-1} + ar^n$$

where the second equation is obtained by multiplying the first by r. Subtract the second equation from the first.

$$S_n - rS_n = a - ar^n$$

Solve for S_n to obtain

$$S_n = \frac{a(1 - r^n)}{1 - r}$$

To show that a sum exists for an infinite series, consider the sequence of partial sums, S_n, defined by

$$S_n = a_1 + a_2 + a_3 + \ldots + a_n, \quad (n = 1, 2, 3, \ldots)$$

If S_n approaches a limit S as n approaches infinity, then S is defined to be the sum of the infinite series and the series is said to **converge to the sum** S. If S_n does not approach a limit, then the series is said to be **divergent**.
Examples:

    a) $1 + 1 + 1 + 1 + \ldots$ is divergent.
    b) $1 - 1 + 1 - 1 + 1 - \ldots$ is divergent.

c) $\dfrac{1}{2} + \dfrac{1}{4} + \ldots + \dfrac{1}{2n} + \ldots$ is convergent.

Consider the infinite geometric series

$$a + ar + ar^2 + \ldots + ar^n + \ldots$$

with $|r| < 1$, the partial sum of which is

$$S_n = \frac{a(1 - r^n)}{1 - r} = \frac{a}{1 - r} - \frac{ar^n}{1 - r}$$

Since $|r| < 1$, $r^n \to 0$ as n increases indefinitely, and the second fraction becomes 0 in the limit. We then have

$$S = \frac{a}{1 - r}$$

and the series converges.

86. Tests for Convergence. Although it is not always possible to find the sum of an infinite series, we can establish certain tests to prove whether or not the sum exists. We first assume the following properties of limits:

1. $\lim\limits_{x \to a} (u + v - w) = \lim\limits_{x \to a} u + \lim\limits_{x \to a} v - \lim\limits_{x \to a} w$

2. $\lim\limits_{x \to a} (uv) = \left(\lim\limits_{x \to a} u\right) \left(\lim\limits_{x \to a} v\right)$

3. $\lim\limits_{x \to a} \left(\dfrac{u}{v}\right) = \dfrac{\lim\limits_{x \to a} u}{\lim\limits_{x \to a} v}$

Theorem. *The series*

$$a_1 + a_2 + a_3 + \ldots + a_n + \ldots$$

cannot converge unless

$$\lim_{n \to \infty} a_n = 0$$

Proof. If the series converges, we must have

$$\lim_{n \to \infty} S_n = S = \lim_{n \to \infty} S_{n-1}$$

and since

$$S_n = a_1 + a_2 + \ldots + a_{n-1} + a_n$$

$$S_{n-1} = a_1 + a_2 + \ldots + a_{n-1}$$

we have by a subtraction

$$S_n - S_{n-1} = a_n$$

Therefore

$$\lim_{n \to \infty} a_n = \lim_{n \to \infty} (S_n - S_{n-1})$$

$$= \lim_{n \to \infty} S_n - \lim_{n \to \infty} S_{n-1}$$

$$= S - S = 0$$

This is an excellent example of a theorem which states a necessary but not a sufficient condition. In fact there are divergent series for which $\lim_{n \to \infty} a_n = 0$; however, there are no convergent series for which this limit is not 0.

Illustration.   Show that the series

$$\frac{1}{2} + \frac{2}{3} + \frac{3}{4} + \ldots + \frac{n}{n+1} + \ldots$$

is divergent.

Solution.   Write

$$a_n = \frac{n}{n+1} = \frac{1}{1 + \dfrac{1}{n}}$$

Then

$$\lim_{n \to \infty} a_n = \lim_{n \to \infty} \left(\frac{1}{1 + \dfrac{1}{n}} \right) = 1 \neq 0$$

Therefore the series cannot converge; i.e., it is divergent.

The Ratio Test.   Let

$$a_1 + a_2 + a_3 + \ldots + a_n + a_{n+1} + \ldots$$

be an infinite series of positive terms and let

$$\lim_{n \to \infty} \left(\frac{a_{n+1}}{a_n} \right) = R$$

Then if

$R < 1$, the series is convergent,
$R > 1$, the series is divergent,
$R = 1$, the test fails.

Illustration.   Test the following for convergence:

$$1 + \frac{1}{2!} + \frac{1}{3!} + \ldots \frac{1}{n!} + \ldots$$

(see p. 182 for definition of $n!$).
Solution. $a_n = \dfrac{1}{n!}$ and $a_{n+1} = \dfrac{1}{(n+1)!}$

The ratio

$$\frac{a_{n+1}}{a_n} = \frac{\dfrac{1}{(n+1)!}}{\dfrac{1}{n!}} = \frac{n!}{(n+1)!}$$

$$= \frac{n!}{(n+1)n!} = \frac{1}{n+1}$$

$$\lim_{n \to \infty} \frac{a_{n+1}}{a_n} = 0 < 1$$

$$\therefore \text{Convergent}$$

The Comparison Test. A series of positive terms converges if each term is less than or equal to the corresponding term of a known convergent series of positive terms.

This is an ideal test to use providing we have a good repertory of series to compare with. We shall discuss some of the well-known series in the next sections.

Illustration. Test the series for convergence:

$$1 + \sum_{i=1}^{\infty} \frac{1}{i} 2^{-i}$$

Solution. Compare with the series

$$1 + \frac{1}{2} + \frac{1}{2^2} + \frac{1}{2^3} + \ldots + \frac{1}{2^n}$$

The given series is

$$1 + \frac{1}{1 \cdot 2} + \frac{1}{2 \cdot 2^2} + \frac{1}{3 \cdot 2^3} + \ldots$$

A comparison reveals

$$1 = 1, \frac{1}{2} = \frac{1}{2}, \frac{1}{2 \cdot 2^2} < \frac{1}{2^2}, \ldots, \frac{1}{n \cdot 2^n} < \frac{1}{2^n}, \ldots$$

Since the given series is term for term less than or equal to the G.P. with $a = 1, r = 1/2$, which is known to converge, the given series converges.

A series whose terms are alternately positive and negative is called an **alternating series.** An alternating series converges if $\lim_{n \to \infty} a_n = 0$ and $|a_{n+1}| < |a_n|$ for all values of n.

87. Some Special Series. We shall now display a few interesting series which may be used in the comparison test. We have already discussed the first two.

I. *The arithmetic series.*

$$a + (a + d) + (a + 2d) + \ldots + [a + (n - 1)d] + \ldots$$

is divergent.

II. *The geometric series.*

$$a + ar + ar^2 + ar^3 + \ldots + ar^n + \ldots$$

is convergent if $|r| < 1$ and divergent for all other values of r.

III. *The p-series.* Let us consider the series

$$1 + \frac{1}{2^p} + \frac{1}{3^p} + \frac{1}{4^p} + \ldots + \frac{1}{n^p} + \ldots$$

a) $p > 1$. We shall compare it with the geometric series with $a = 1$ and $r = 2^{-(p-1)}$

$$1 = 1$$

$$\frac{1}{2^p} + \frac{1}{3^p} < \frac{1}{2^p} + \frac{1}{2^p} = \frac{2}{2^p} = \frac{1}{2^{p-1}}$$

$$\frac{1}{4^p} + \frac{1}{5^p} + \frac{1}{6^p} + \frac{1}{7^p} < \frac{1}{4^p} + \frac{1}{4^p} + \frac{1}{4^p} + \frac{1}{4^p} = \frac{4}{4^p} = \left(\frac{1}{2^{p-1}}\right)^2$$

$$\frac{1}{8^p} + \frac{1}{9^p} + \ldots + \frac{1}{15^p} < \left(\frac{1}{2^{p-1}}\right)^3$$

etc.

Since $|r| < 1$, the geometric series converges and so does the p-series.

b) $p = 1$. When $p = 1$ we have a special series called the **harmonic series.**

$$1 + \frac{1}{2} + \frac{1}{3} + \frac{1}{4} + \ldots + \frac{1}{n} + \ldots$$

Set up the following comparison:

$$1 + \frac{1}{2} = \frac{3}{2} > \frac{1}{2}$$

$$\frac{1}{3} + \frac{1}{4} = \frac{7}{12} > \frac{1}{2}$$

$$\frac{1}{5} + \frac{1}{6} + \frac{1}{7} + \frac{1}{8} > \frac{1}{8} + \frac{1}{8} + \frac{1}{8} + \frac{1}{8} = \frac{4}{8} = \frac{1}{2}$$

$$\frac{1}{9} + \frac{1}{10} + \ldots + \frac{1}{16} > \frac{1}{16} + \frac{1}{16} + \ldots + \frac{1}{16} = \frac{8}{16} = \frac{1}{2}$$

We see that the series is greater than the series

$$\frac{1}{2} + \frac{1}{2} + \frac{1}{2} + \ldots = \frac{1}{2}(1 + 1 + 1 + \ldots)$$

which is divergent.

c) $p < 1$. In this case each term is greater than those of the harmonic series; e.g.,

$$\frac{1}{\sqrt{2}} > \frac{1}{2}$$

Therefore the series is divergent for $p < 1$.

IV. *The telescopic series.* The following series is referred to as a telescopic series:

$$\frac{1}{k(k + 1)} + \frac{1}{(k + 1)(k + 2)} + \ldots + \frac{1}{(k + n - 1)(k + n)} + \ldots$$

where $k > 0$. The general term may be written in the form (see Section 73)

$$\frac{1}{(k + n - 1)(k + n)} = \frac{1}{k + n - 1} - \frac{1}{k + n}$$

The sequence of partial sums is

$$S_1 = \frac{1}{k} - \frac{1}{k + 1}$$

$$S_2 = \frac{1}{k} - \frac{1}{k + 1} + \frac{1}{k + 1} - \frac{1}{k + 2} = \frac{1}{k} - \frac{1}{k + 2}$$

$$S_3 = \frac{1}{k} - \frac{1}{k + 2} + \frac{1}{k + 2} - \frac{1}{k + 3} = \frac{1}{k} - \frac{1}{k + 3}$$

$$\cdot \quad \cdot \quad \cdot \quad \cdot \quad \cdot \quad \cdot \quad \cdot \quad \cdot$$

$$S_n = \frac{1}{k} - \frac{1}{k + n}$$

Now the limit of S_n as n increases is

$$\lim_{n \to \infty} S_n = \lim_{n \to \infty} \left(\frac{1}{k} - \frac{1}{k + n} \right) = \frac{1}{k}$$

Consequently the telescopic series is convergent; in fact, the sum is $1/k$.

Illustrations.   Test the following for convergence:

1) $1 + \dfrac{1}{2^2} + \dfrac{1}{3^3} + \dfrac{1}{4^4} + \ldots + \dfrac{1}{n^n} + \ldots$

Solution.   Compare with the p-series:

$$1 = 1, \frac{1}{2^2} = \frac{1}{2^2}, \frac{1}{3^3} < \frac{1}{3^2}, \ldots, \frac{1}{n^n} < \frac{1}{n^2}$$

The series is term for term less than the p-series with $p = 2$ and, therefore, converges.

2) $1 + \dfrac{1}{2}\sqrt{2} + \dfrac{1}{3}\sqrt{3} + \ldots + \dfrac{1}{n}\sqrt{n} + \ldots$

Solution.   The series may be written in the form

$$1 + \frac{1}{\sqrt{2}} + \frac{1}{\sqrt{3}} + \ldots + \frac{1}{\sqrt{n}}$$

which is the p-series with $p = 1/2$. Thus the series is divergent.

88. Binomial Series. The raising of a binomial to an integral power can be expressed by the **binomial formula**

$$(a + b)^n = a^n + na^{n-1}b + \frac{n(n-1)}{1\cdot2}a^{n-2}b^2 + \frac{n(n-1)(n-2)}{1\cdot2\cdot3}a^{n-3}b^3$$
$$+ \ldots + nab^{n-1} + b^n$$

Example:
$$(a + b)^5 = a^5 + 5a^4b + \frac{(5)(4)}{2}a^3b^2 + \frac{(5)(4)(3)}{(2)(3)}a^2b^3$$
$$+ \frac{(5)(4)(3)(2)}{(2)(3)(4)}ab^4 + b^5$$

$$= a^5 + 5a^4b + 10a^3b^2 + 10a^2b^3 + 5ab^4 + b^5$$

This is a finite series if n is a positive integer, and the general formula can be established by mathematical induction. The general (rth) term has the form

$$\frac{n(n-1)(n-2)\ldots(n-r+2)}{1\cdot2\cdot3\cdot\ldots(r-1)}a^{n-r+1}b^{r-1}$$

The denominator is the product of the positive integers from 1 to

$(r - 1)$. This product occurs with sufficient frequency to be given a special notation.

Definition.   *The product of all the positive integers from* 1 *to* n, *inclusive, is denoted by the symbol* **n**!

$$1 \cdot 2 \cdot 3 \cdot 4 \cdot \ldots n = n!$$

It is read "n factorial." The values of factorials for $1 \le n \le 10$ are given in the following table:

n	$n!$	n	$n!$
1	1	6	720
2	2	7	5040
3	6	8	40320
4	24	9	362880
5	120	10	3628800

Factorials have many interesting properties of which the following two can be obtained directly from the definition.

1. $n! = n \cdot (n - 1)!$

2. If $r < n$, then $\dfrac{n!}{r!} = n(n - 1) \ldots (r + 1)$.

We shall also define $0! = 1$. This is logical since if $n = 1$ in the first property, we have

$$1! = 1 \cdot 0! \quad \text{or} \quad 0! = 1$$

Let us now consider the binomial series when n is any rational number which is not a positive integer. The resulting series is an infinite series.

Examples:

a) $(x^2 - y)^{\frac{1}{2}} = (x^2)^{\frac{1}{2}} + \dfrac{1}{2}(x^2)^{-\frac{1}{2}}(-y) + \dfrac{(\frac{1}{2})(-\frac{1}{2})}{2!}(x^2)^{-\frac{3}{2}}(-y)^2$

$\qquad + \dfrac{(\frac{1}{2})(-\frac{1}{2})(-\frac{3}{2})}{3!}(x^2)^{-\frac{5}{2}}(-y)^3 + \ldots$

$\qquad = x - \dfrac{1}{2}x^{-1}y - \dfrac{1}{8}x^{-3}y^2 - \dfrac{1}{16}x^{-5}y^3 - \ldots$

b) $\dfrac{1}{1 + x} = (1 + x)^{-1} = 1^{-1} + (-1)(1)^{-2}x + \dfrac{(-1)(-2)}{2!}(1)^{-3}x^2$

$\qquad + \dfrac{(-1)(-2)(-3)}{3!}(1)^{-4}x^3 + \ldots$

$$= 1 - x + x^2 - x^3 + \ldots$$

89. Power Series.

A series of the form

$$a_0 + a_1 x + a_2 x^2 + \ldots + a_n x^n + \ldots$$

where x is a variable and a_i $(i = 0, 1, 2, \ldots)$ are constants is called a **power series in x.** If there exist values of x for which the series converges, then for each such value of x, the sum can be determined. We can denote the sum by y. Thus the power series in x defines a function y over the domain of values of x for which the series converges; the domain is called the **interval of convergence.** The last example of Section 88 is an illustration of a power series developed by the use of the binomial formula.

We shall explain a procedure for determining the interval of convergence by considering an illustration.

Illustration. Determine the interval of convergence for the power series

$$1 + x + \frac{x^2}{2} + \frac{x^3}{3} + \ldots + \frac{x^n}{n} + \ldots$$

Solution. Consider the ratio

$$a_{n+1} \div a_n = \frac{x^{n+1}}{n+1} \div \frac{x^n}{n} = x\left(\frac{n}{n+1}\right)$$

The limit of this ratio as $n \to \infty$ is

$$R = \lim_{n \to \infty} x\left(\frac{n}{n+1}\right) = x$$

since

$$\lim_{n \to \infty} \left(\frac{n}{n+1}\right) = \lim_{n \to \infty} \left(\frac{1}{1 + \frac{1}{n}}\right) = 1$$

By the ratio test the series converges for $|R| = |x| < 1$, so that the interval of convergence is $-1 < x < 1$. However, the test fails if $R = 1 = |x|$. We need to test the series for these two values of x. If $x = 1$, the series is

$$1 + 1 + \frac{1}{2} + \frac{1}{3} + \frac{1}{4} + \ldots + \frac{1}{n} + \ldots$$

which is the divergent harmonic series. If $x = -1$, the series is

$$1 - 1 + \frac{1}{2} - \frac{1}{3} + \frac{1}{4} + \ldots + (-1)^n \left(\frac{1}{n}\right) + \ldots$$

which is an alternating series for which $\lim_{n \to \infty} a_n = 0$ and $a_{n+1} < a_n$ and thus is convergent. The interval of convergence is

$$[-1, 1[\, = \, -1 \le x < 1$$

There are other methods besides the binomial formula for generating power series. Most formulas depend upon a knowledge of the calculus for their derivation.* These may be applied to give series representation for functions. We shall list three of those that are well known.

I. *Reciprocal series*. These can be obtained from the binomial series. The interval of convergence is also given.

$$\frac{1}{1-x} = 1 + x + x^2 + x^3 + \ldots + x^n + \ldots, \quad (-1 < x < 1)$$

$$\frac{1}{1+x} = 1 - x + x^2 - x^3 + \ldots, \quad (-1 < x < 1)$$

II. *Exponential series*. The series is given by

$$e^x = 1 + x + \frac{x^2}{2!} + \frac{x^3}{3!} + \ldots + \frac{x^n}{n!} + \ldots$$

$$= \sum_{n=0}^{\infty} \frac{x^n}{n!}, \quad \text{(all } x\text{)}$$

III. *Logarithmic series*. The series is expressed by considering $\ln (1 + x)$.

$$\ln (1 + x) = x - \frac{x^2}{2} + \frac{x^3}{3} - \frac{x^4}{4} + \ldots, \quad (-1 < x \le 1)$$

Series are used to calculate values of the functions for given values of the argument.

Illustration.   Find $\ln 1.01$ to seven decimal places.
Solution. $\ln 1.01 = \ln (1 + .01)$ and $x = .01$

$$\ln (1 + .01) = (.01) - \frac{(.01)^2}{2} + \frac{(.01)^3}{3} - \frac{(.01)^4}{4} + \ldots$$

$$= .01 - .00005 + .00000033 - .0000000025$$

$$= .0099503$$

90. Operations with Series.   We may add, subtract, multiply, and divide series to obtain other series. The resultant series are usually convergent over the intersection of the intervals of the given series.

---

*See C. O. Oakley, *The Calculus,* New York: Barnes and Noble, Inc., 1957.

Illustration. Use the exponential series to find the series for cosh x.
Solution.

$$e^x = 1 + x + \frac{x^2}{2!} + \frac{x^3}{3!} + \ldots + \frac{x^n}{n!} + \ldots$$

$$e^{-x} = 1 - x + \frac{x^2}{2!} - \frac{x^3}{3!} + \ldots + (-1)^n \frac{x^n}{n!} + \ldots$$

Add and divide by 2 to obtain

$$\frac{e^x + e^{-x}}{2} = 1 + \frac{x^2}{2!} + \frac{x^4}{4!} + \ldots + \frac{x^{2n}}{(2n)!} + \ldots = \cosh x$$

Summary of Symbols	
A.P., arithmetic progression G.P., geometric progression $\lim\limits_{n \to \infty}$, limit as n approaches infinity	$\sum\limits_{i=1}^{n} a_i$, the sum of a_i from $i = 1$ to $i = n$ $n!$, n factorial
$a_1 + a_2 + a_3 + \ldots + a_n + \ldots$, a series of constants $a_0 + a_1 x + \ldots + a_n x^n + \ldots$, a power series	

91. Exercise X.

1. Assuming that the numbers form either an A.P. or a G.P., continue the following sequences for three additional terms:
   a) $3, 1, -1, -3, \ldots$ b) $-7, -4, -1, \ldots$
   c) $7, 21, 63, \ldots$ d) $-2, 4, -8, \ldots$
   e) $9, 13, 17, \ldots$ f) $32, 16, 8, 4, \ldots$
2. Again referring to either an A.P. or a G.P., find the missing parts:
   a) Given $a = 3, d = 7, n = 9$; find l and S.
   b) Given $a = 3, d = 6, S = 363$; find n and l.
   c) Given $a = -8, l = 8, n = 5$; find d and S.
   d) $a = 12, r = 1/2, n = 7$; find l and S.
   e) $a = 27/4, r = 2/3, l = 8/9$; find n and S.
   f) $a = 3, n = 3, S = 63$; find l and r.
   g) Insert five arithmetic means between 63 and 9.
   h) Insert two geometric means between 2 and 1024.
3. Find the sums given by the following:
   a) $\sum\limits_{k=0}^{n} 2^k$

   b) $1 + 3 + 5 + \ldots + (2n - 1)$

   c) $2 + 4 + 6 + \ldots + 2n$

4. Find the sum of the following infinite geometric series:

   a) $1 + \dfrac{1}{2} + \dfrac{1}{4} + \ldots + \dfrac{1}{2^{n-1}} + \ldots$

   b) $1 - \dfrac{1}{2} + \dfrac{1}{4} - \ldots + (-1)^{n-1} \dfrac{1}{2^{n-1}} + \ldots$

   c) $\displaystyle\sum_{n=1}^{\infty} \left(\dfrac{1}{3^{n-1}} \right)$ d) $\displaystyle\sum_{n=1}^{\infty} \left(\dfrac{1}{5^{n-1}} \right)$

   e) $\displaystyle\sum_{n=1}^{\infty} 6 \left(\dfrac{2}{3} \right)^{n-1}$ f) $\displaystyle\sum_{n=1}^{\infty} \left(\dfrac{1}{10} \right)^{n-1}$

   g) Write out the series for (f) above and put into decimal form.

5. Test the following series for convergence and divergence:

   a) $1 + \dfrac{1}{2!} + \dfrac{1}{3!} + \dfrac{1}{4!} + \ldots$

   b) $1 + \dfrac{1}{2} + \dfrac{2}{3} + \dfrac{3}{4} + \dfrac{4}{5} + \ldots$

   c) $\dfrac{1 \cdot 2}{3} + \dfrac{2 \cdot 3}{4} + \dfrac{3 \cdot 4}{5} + \ldots + \dfrac{n(n+1)}{n+2} + \ldots$

   d) $\dfrac{3}{1 \cdot 2} + \dfrac{4}{2 \cdot 3} + \dfrac{5}{3 \cdot 4} + \ldots + \dfrac{n+2}{n(n+1)} + \ldots$

   e) $\dfrac{1}{3} + \dfrac{3}{5} + \dfrac{5}{7} + \dfrac{7}{9} + \ldots$

   f) $1 - \dfrac{1}{3} + \dfrac{1}{5} - \dfrac{1}{7} + \dfrac{1}{9} - \ldots$

   g) $2 + \dfrac{2 \cdot 4}{2!} + \dfrac{2 \cdot 4 \cdot 6}{3!} + \dfrac{2 \cdot 4 \cdot 6 \cdot 8}{4!} + \ldots + \dfrac{2 \cdot 4 \cdots (2n)}{n!}$

6. Find the first four terms in the expansion of each of the following binomials:

   a) $(x - 2y^2)^7$ b) $(3x - 2y)^{\frac{1}{3}}$

   c) $\left(x^{\frac{1}{2}} - 2y^{\frac{1}{3}} \right)^{-2}$ d) $(a + bx)^7$

e) $(1 + x)^{\frac{1}{2}}$ f) $(1 + x)^{-\frac{1}{2}}$

g) $(2 + 3x)^{-3}$ h) $(1 + bx)^{-2}$

7. Find the indicated term of the binomial expansion:

   a) 7th term of $\left(2x - 3y^{\frac{1}{2}}\right)^{-\frac{1}{2}}$

   b) 9th term of $(x - y)^{15}$

   c) 6th term of $(a + bx)^{-2}$

8. Find the interval of convergence for each of the following power series:

   a) Parts (e), (f), (g), and (h) of Exercise 6, above.

   b) $1 + \dfrac{x}{2} + \dfrac{2x^2}{3} + \dfrac{3x^3}{4} + \ldots$

   c) $x - \dfrac{x^3}{3} + \dfrac{x^5}{5} - \dfrac{x^7}{7} + \ldots$

   d) $1 - \dfrac{x^2}{2!} + \dfrac{x^4}{4!} - \dfrac{x^6}{6!} + \ldots$

   e) $1 + (x - 1) + \dfrac{(x - 1)^2}{2} + \dfrac{(x - 1)^3}{3} + \ldots$

9. Find the series for $\sinh x$ and its interval of convergence.

10. Use a series expansion to find

$$\begin{vmatrix} 1 & x \\ x & 1 \end{vmatrix}^{\frac{1}{2}}$$

and find its value for $x = 1/2$ to three decimal places.

IX

PARTITIONS, PERMUTATIONS, AND PROBABILITY

92. Definitions.   The definition of disjoint sets was given in Section 21. We shall now need the following definitions.

Definition.   *Two subsets* A *and* B *of a set* S *are said to be* **exhaustive** *if and only if their union is* S.

Examples:

a) $A = \{a, b\}$ and $B = \{c, d, e\}$ are exhaustive subsets of $S = \{a, b, c, d, e\}$.

b) $[2, 3]$ and $]3, 5]$ are exhaustive subsets of $[2, 5]$.

c) $[2, 3[$ and $]3, 5]$ are *not* exhaustive subsets of $[2, 5]$ since 3 is not included in either subset.

Definition.   *A* **partition** *of a set* S *is a subdivision of the set into subsets that are disjoint and exhaustive.*

Examples: Let $S = \{a, b, c, d, e\}$. Then the following are partitions of S.

a) $A_1 = \{a, b\}$ and $A_2 = \{c, d, e\}$.

b) $A_1 = \{a, b\}$, $A_2 = \{c, d\}$, and $A_3 = \{e\}$.

c) $A_1 = \{a\}$, $A_2 = \{b\}$, $A_3 = \{c\}$, $A_4 = \{d\}$, and $A_5 = \{e\}$.

Definition.   *The subsets* A_i *of* S *which form a partition are called* **cells** *of the partition.*

If A_i and B_i are two partitions of the same set S, then the collection of sets $A_i \cap B_i$ is a new partition and is called the **cross-partition** of the original two partitions.

Examples:

a) Let $S = \{a, b, c, d, e\}$ be partitional into $A_1 = \{a, b\}$, $A_2 = \{c, d\}$, $A_3 = \{e\}$, and $B_1 = \{a, b, c\}$, $B_2 = \{d, e\}$. Then

$$A_1 \cap B_1 = \{a, b\} = C_1 \qquad A_2 \cap B_2 = \{d\} = C_4$$

188

$$A_1 \cap B_2 = \quad \emptyset \quad = C_2 \qquad A_3 \cap B_1 = \quad \emptyset \quad = C_5$$

$$A_2 \cap B_1 = \{c\} = C_3 \qquad A_3 \cap B_2 = \{e\} = C_6$$

The collection of subsets C_i is a new partition of S.

The above example shows that some of the cells of a partition may be empty. This is permissible since empty sets do not violate the criteria for a partition; i.e., the intersection of two empty sets is empty and therefore they are disjoint.

b) Let S be the set of real numbers in the closed interval $[1, 5]$. Let the set be partitioned in two ways, $A_1 = [1, 3[$, $A_2 = [3, 5]$, and $B_1 = [1, 2[$, $B_2 = [2, 3]$, $B_3 =]3, 4[$, $B_4 = [4, 5]$. The cross-partition is the following collection of subsets:

$$C_1 = A_1 \cap B_1 = [1, 2[\qquad C_2 = A_1 \cap B_2 = [2, 3[$$

$$C_3 = A_1 \cap B_3 = \emptyset \qquad C_4 = A_1 \cap B_4 = \emptyset$$

$$C_5 = A_2 \cap B_1 = \emptyset \qquad C_6 = A_2 \cap B_2 = \{3\}$$

$$C_7 = A_2 \cap B_3 =]3, 4[\qquad C_8 = A_2 \cap B_4 = [4, 5]$$

Notice that in the two examples the number of cells of the cross-partitions is equal to the product of the number of cells in the two given partitions. In the first example we considered a set S with a finite number of elements. In the second example the set S had an infinite number of elements.

Definition.   *The symbol* n(A) *is used to denote the number of elements in the set* A.

Example: If $A = \{a, b, c\}$, then $n(A) = 3$.

Property 1.   *If* A *and* B *are disjoint sets, then*

$$n(A \cup B) = n(A) + n(B)$$

This property follows directly from the definitions.

Property 2.   *If* A *and* B *are not disjoint, then*

$$n(A \cup B) = n(A) + n(B) - n(A \cap B)$$

Proof.   We can divide A into two disjoint sets $A \cap B'$ and $A \cap B$, since B and B' are disjoint. (B' is the complement of B.) Similarly, B can be divided into $B \cap A'$ and $B \cap A$, which are disjoint. Then by Property 1

$$n(A) = n(A \cap B') + n(A \cap B)$$

$$n(B) = n(A' \cap B) + n(A \cap B)$$

Adding, we obtain

$$n(A) + n(B) = n(A \cap B') + n(A' \cap B) + 2n(A \cap B)$$

Sets $A \cap B'$, $A' \cap B$, and $A \cap B$ are three disjoint sets whose union is $A \cup B$, so that by Property 1

$$n(A \cap B') + n(A' \cap B) + n(A \cap B) = n(A \cup B)$$

If we substitute this into the above equation we have

$$n(A) + n(B) = n(A \cup B) + n(A \cap B)$$

or

$$n(A \cup B) = n(A) + n(B) - n(A \cap B)$$

Example: If $A = \{a, b, c, d\}$ and $B = \{b, d, e\}$, then $A \cup B = \{a, b, c, d, e\}$ and $A \cap B = \{b, d\}$. The number of elements is $n(A) = 4$, $n(B) = 3$, $n(A \cap B) = 2$, and $n(A \cup B) = 4 + 3 - 2 = 5$.

For three or more sets the formulas for the number of elements of the unions become cumbersome. Instead we use Venn diagrams.

Illustrations.

1) Find the number of elements in the union of sets A, B, and C if $n(A) = 16$, $n(B) = 12$, $n(C) = 15$, $n(A \cap B) = 5$, $n(A \cap C) = 3$, $n(B \cap C) = 7$, and $n(A \cap B \cap C) = 2$.

Solution. Draw the Venn diagram shown in Figure 53. Start by placing the numbers in the appropriate intersections, making sure that the totals check the given data. The number of elements in the union is the total of all the numbers inside the circles, which is 30.

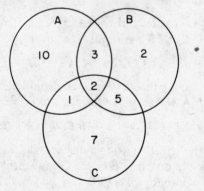

Fig. 53

2) In a reading survey the following data were collected concerning the reading of three magazines A, B, and C.

A	B	C	A & B	A & C	B & C	A & B & C
50%	45%	55%	20%	15%	25%	5%

a) What per cent read at least two magazines?
b) What per cent read exactly two magazines?
c) What per cent do not read any of the three?

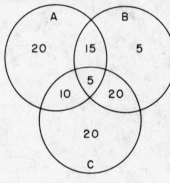

Fig. 54

Solution. Draw the Venn diagram shown in Figure 54. By adding the percentages shown we have the following: (a) 50%, (b) 45%, (c) 100%— 95% = 5%.

93. Permutations. We first considered the idea of ordering the elements of a set when we defined a sequence. Let us extend the concept a little further.

Fundamental Principle. *If one thing can be done in* p *different ways and after it has been done in any one of these ways, a second thing can be done in* q *different ways, then the two things can be done together, in the order stated, in* pq *different ways.*

The principle can be generalized to a third thing in *r* different ways, a fourth thing in *s* different ways, etc., and the number of different ways in which all things can be done in the order stated is

$$p \cdot q \cdot r \cdot s \cdots$$

Example: Given three objects *a*, *b*, and *c*. It is desired to put them into three baskets. There are three ways in which an object can be put in the first basket; namely, either *a*, or *b*, or *c*. Having put an object in the first basket, there are two objects left to put into the second basket, so that for each of the three ways there are now two choices. Having placed objects in the first two baskets there is only one object left for the third basket. Hence there are $(3)(2)(1) = 6$ ways in which the objects can be placed in the baskets.

The graphical representation of the above procedure is referred to as a tree graph. We shall place the tree upside down. See Figure 55.

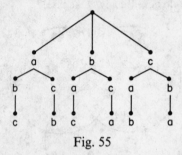

Fig. 55

The number of elements in the Cartesian product of two finite sets is the product of the number of elements in each set; i.e.,

$$n(A \times B) = n(A) \cdot n(B)$$

Example: If $A = \{2, 3, 5\}$ and $B = \{1, 4, 6\}$, then $A \times B = \{(2, 1), (2, 4), (2, 6), (3, 1), (3, 4), (3, 6), (5, 1), (5, 4), (5, 6)\}$ and $n(A \times B) = n(A) \cdot n(B) = (3)(3) = 9$.

Definitions.   *An* **arrangement** *is a selection in a definite order. Each different arrangement which can be made from a given number of objects by taking any part or all of them at a time is called a* **permutation**.

The symbol $P(n, r)$ is used to denote the *number of permutations of* **n** *things taken* **r** *at a time.* The following formulas are based on the fundamental principle.

   I. The permutation of *n different* things taken *r* at a time is

$$P(n, r) = n(n - 1)(n - 2) \ldots (n - r + 1)$$

$$= \frac{n!}{(n - r)!}$$

   II. The permutation of *n* different things taken *n* at a time is

$$P(n, n) = n!$$

III. The number of distinct permutations of n things taken all at a time of which p are alike, q others are alike, r others are alike, etc., is

$$P = \frac{n!}{p!q!r!\ldots}$$

Examples:
a) $P(8, 2) = 8 \cdot 7 = 56$
b) $P(30, 3) = 30 \cdot 29 \cdot 28 = 24360$
c) $P(7, 7) = 7! = 5040$.

Illustrations.

1) Eight swimmers compete in the finals of a meet in which only the first three places receive medals. In how many ways can the medals be distributed?
Solution. $P(8, 3) = 8 \cdot 7 \cdot 6 = 336$

2) How many permutations can be made of the letters in the word *Mississippi* when taken all at a time?
Solution. This is a permutation of 11 things taking all of them at a time. Since there are 4s's, 4i's, and 2p's, we have

$$\frac{11!}{4!4!2!} = 34650$$

We shall consider two special kinds of petmutations. A **cyclical permutation** is an arrangement of n elements in which the nth element is adjacent to the first. This will cause the same arrangement to occur n times. To obtain the different arrangements we fix one element and permute the other $n - 1$ elements. Thus the number of cyclical permutations that can be formed from n elements of a set is $(n - 1)!$
Example: The number of ways four people can be seated at a round table is $(n - 1)! = 3! = 6$.

Suppose in the formation of an arrangement we permit the same element to be used more than once; i.e., we permit repetitions. From a set of n elements there are n available for the first position. Since this element can be used again, there are still n elements available for the second position. The same reasoning holds for all the positions. The fundamental principal states that from a set of n elements there are $n \cdot n \cdot n \cdot \ldots \cdot n = n^r$ arrangements of r elements with repetitions.
Example: The number of arrangements of $\{a, b, c\}$ using two elements with repetition permitted is $3^2 = 9$. They are aa, ab, ac, bb, ba, bc, cc, ca, cb. Draw the tree graph for this example.

94. Combinations. We turn to the problem of counting the number of different *nonordered* subsets each containing r elements and formed from a set of n elements ($r \leq n$). In this case we do not consider the subsets $\{a, b\}$ and $\{b, a\}$ as different. Thus there should be fewer nonordered subsets.

Definition.   *All the possible* **selections** *consisting of* **r** *different things chosen from* **n** *given things* ($r \leq n$), *without regard to the order of selection, are called* **combinations.**

We use the symbol $C(n, r)$, which is read, "the number of combinations of n things taken r at a time." Since each of the $C(n, r)$ combinations contains r elements which can be arranged in $r!$ permutations, we see that if we multiply $C(n, r)$ by $r!$ we obtain the total number of permutations that can be formed from the n elements. Thus we have

$$r!\, C(n, r) = P(n, r) = \frac{n!}{(n - r)!}$$

or, solving for $C(n, r)$,

$$C(n, r) = \frac{n!}{r!(n - r)!}$$

Examples:

a) $C(7, 3) = \dfrac{7!}{3!\,4!} = \dfrac{7 \cdot 6 \cdot 5}{1 \cdot 2 \cdot 3} = 35$

b) $C(12, 8) = \dfrac{12!}{8!\,4!} = \dfrac{12 \cdot 11 \cdot 10 \cdot 9}{1 \cdot 2 \cdot 3 \cdot 4} = 495$

c) $C(40, 2) = \dfrac{40!}{2!\,38!} = \dfrac{40 \cdot 39}{1 \cdot 2} = 780$

Illustration.   In how many ways can a customer purchase 3 magazines from a stand containing 22 different magazines?
Solution.   This is a selection of 3 from 22.

$$C(22, 3) = \frac{22 \cdot 21 \cdot 20}{1 \cdot 2 \cdot 3} = 1540$$

Let us consider the binomial expansion of

$$(1 + x)^n = 1 + nx + \frac{n(n - 1)}{1 \cdot 2} x^2 + \frac{n(n - 1)(n - 2)}{1 \cdot 2 \cdot 3} x^3$$

$$+ \frac{n(n - 1)(n - 2)(n - 3)}{1 \cdot 2 \cdot 3 \cdot 4} x^4 + \ldots + x^n$$

$$= a_0 + a_1 x + a_2 x^2 + a_3 x^3 + a_4 x^4 + \ldots + a_n x^n$$

Note that

$$a_1 = n = C(n, 1)$$

$$a_2 = \frac{n(n-1)}{1 \cdot 2} = C(n, 2)$$

$$a_3 = \frac{n(n-1)(n-2)}{1 \cdot 2 \cdot 3} = C(n, 3)$$

$$\ldots$$

$$a_n = 1 = C(n, n)$$

Thus the coefficients of a binomial expansion can be written in terms of the combinations symbol. In fact, there are three ways to write the *binomial coefficients*:

$$C(n, k) = \binom{n}{k} = \frac{n(n-1)\ldots(n-k+1)}{k!}$$

Let $x = 1$ in the expansion; then

$$(1 + 1)^n = 1 + C(n, 1) + C(n, 2) + C(n, 3) + \ldots + C(n, n)$$

Combine the numerical values to obtain

$$C(n, 1) + C(n, 2) + \ldots + C(n, n) = 2^n - 1$$

This is the total number of combinations of n things taken *successively* 1, 2, 3, ..., n at a time.

> **Illustration.** How many different sums of money can be formed from a nickel, a dime, and a quarter?
>
> *Solution.* We have a set of 3 elements from which to select successively 1, 2, and 3 at a time. Since $n = 3$ we have
>
> $$2^n - 1 = 2^3 - 1 = 8 - 1 = 7$$

Property 1. $C(n, k) = C(n, n - k)$

Proof. By the formula: $C(n, k) = \dfrac{n!}{k!(n-k)!}$

Since $n - (n - k) = k$: $C(n, n - k) = \dfrac{n!}{(n-k)!k!}$

Property 2. $C(n + 1, k) = C(n, k) + C(n, k - 1)$

Proof.

$$C(n, k) + C(n, k - 1) = \frac{n!}{k!(n-k)!} + \frac{n!}{(k-1)!(n-k+1)!}$$

$$= \frac{n!}{k(k-1)!(n-k)!}$$

$$+ \frac{n!}{(k-1)!(n-k)!(n-k+1)}$$

$$= \frac{n!(n-k+1) + n!(k)}{k!(n+1-k)!}$$

$$= \frac{n!(n+1)}{k!(n+1-k)!}$$

$$= C(n+1, k)$$

Illustrations.

1) Given $C(n, 2) = 66$; find n.
Solution.

$$C(n, 2) = \frac{n(n-1)}{2} = \frac{n^2 - n}{2} = 66$$

$$n^2 - n - 132 = 0 \quad \text{or} \quad (n-12)(n+11) = 0$$

The solution is $n = 12$, since n cannot be negative.

2) Given $C(n, r) = 20$ and $P(n, r) = 120$; find n and r.

Solution. $\qquad C(n, r) = \dfrac{n!}{r!(n-r)!} = 20$

$$P(n, r) = \frac{n!}{(n-r)!} = 120$$

Therefore

$$n! = 20r!(n-r)! = 120(n-r)!$$

or $\qquad\qquad r! = 6 \quad \text{and} \quad r = 3$

We then have

$$P(n, 3) = n(n-1)(n-2) = 120$$

$$n^3 - 3n^2 + 2n - 120 = 0$$

The real positive root is $n = 6$.

95. Mathematical Probability.
The theory of probability, originally developed in connection with games of chance, computations of mortality, and insurance, is now being applied to scientific research and engineering. Suppose an event can result in one of n ways, each of which, in a single trial, is equally likely. Let s of these ways be considered successes and the others, $n - s$, failures. Then the

probability, p, that one particular trial will result in a success is defined to be

$$p = \frac{s}{n}$$

The probability, q, that one trial will result in a failure is defined to be

$$q = \frac{n - s}{n}$$

We see that

$$p + q = \frac{s}{n} + \frac{n - s}{n} = 1$$

$$q = 1 - p$$

$$0 \leq p \leq 1$$

Illustration. What is the probability that one card drawn from a deck of 52 cards will be a heart? An ace? The ace of hearts?

Solution. There are 52 cards, so $n = 52$.

a) There are 13 hearts; thus $s = 13$ and $p = 13/52 = 1/4$.

b) There are 4 aces; thus $s = 4$ and $p = 4/52 = 1/13$.

c) There is one ace of hearts; thus $s = 1$ and $p = 1/52$.

If more than one event is involved, three possibilities present themselves.

Definition. *Two or more events are* **mutually exclusive** *if the occurrence of one excludes the possibility of the occurrence of each of the others.*

Let the mutually exclusive events be $E_1, E_2, \ldots, E_k$ and suppose they can happen in $S_1, S_2, \ldots, S_k$ ways, respectively, in one trial. The total possible result is $n = S_1 + S_2 + \ldots + S_k$. The probability that E_i will happen in one trial is $p_i = S_i/n$, $(i = 1, \ldots, k)$. The probability that some one of them will happen in one trial is

$$p = \frac{S_1}{n} + \frac{S_2}{n} + \frac{S_3}{n} + \ldots + \frac{S_k}{n}$$

$$= p_1 + p_2 + p_3 + \ldots + p_k$$

Illustration. A bag contains 5 black balls, 6 white balls, 4 yellow balls, and 7 red balls. If one ball is drawn from the bag, what is the probability that it will be either black or yellow?

Solution. The events are mutually exclusive, since if a red ball is drawn, then none of the other colors can be drawn in the first draw.

There are 22 balls in the bag, so $n = 22$. The respective probabilities are

$$p_b = \frac{5}{22}, \; p_w = \frac{6}{22}, \; p_y = \frac{4}{22}, \; p_r = \frac{7}{22}$$

The probability that the drawn ball will be either black or yellow is

$$p_b + p_y = \frac{5}{22} + \frac{4}{22} = \frac{9}{22}$$

Definition. *Two or more events are* **independent** *if the occurrence of one does not affect the occurrence of any other. On the other hand, if the occurrence of one event affects the occurrence of other events, the events are said to be* **dependent.**

Example: If in a bag of colored balls, one is withdrawn and then is replaced before a second drawing is made, the events are independent. On the other hand, if the drawn ball is held out, the subsequent events are changed and the events are dependent.

Theorem. *The probability that all of a set of independent events will occur on a given occasion when all of them are possible is the product of their separate probabilities.*

Theorem. *If the probability of a first event is* p_1, *and if, after this has happened, the probability of a second event is* p_2, *then the probability that both events will happen in the order given is* $p_1 p_2$.

Illustrations. Consider a bag containing 5 white balls, 10 black balls, 7 yellow balls, and 13 red balls.

1) Find the probability that on two consecutive draws both balls will be white if the ball is replaced after the first draw.

Solution. There are 35 balls; $n = 35$ and $p_w = 5/35 = 1/7$. The probability that both balls be white is

$$p = \frac{1}{7} \cdot \frac{1}{7} = \frac{1}{49}$$

2) Find the probability that on two consecutive draws both balls will be white if the ball is not replaced after the first draw.

Solution. The probability on the first draw is $p_w = 1/7$. After this draw there are 34 balls left in the bag, of which 4 are white; the probability on the second draw is $p_w = 4/34 = 2/17$. The combined probability is

$$p = \left(\frac{1}{7}\right)\left(\frac{2}{17}\right) = \frac{2}{119}$$

3) If a pair of dice is tossed twice, what is the probability of obtaining a 4 on both tosses?

Solution. There are 6 faces on each die so the number of possibilities for a pair of dice is $n = (6)(6) = 36$. A 4 can be made by (2 and 2), (1 and 3), and (3 and 1); hence it can be thrown in 3 ways. The two tosses are independent; consequently,

$$p = \left(\frac{3}{36}\right)\left(\frac{3}{36}\right) = \frac{1}{144}$$

These concepts of probability can be described very effectively by making use of a few notations from mathematical logic.

Definitions of Symbols

$P(E)$, the probability of event E.

$P(E \wedge F)$, the probability of E and F.

$P(E|F)$, the probability of E after F has occurred.

$P(E \veebar F)$, the probability of E or F but not both.

$P(E \vee F)$, the probability of E or F or both.

We can then summarize the probabilities in the following table.

Events	Probability
Independent	$P(E \wedge F) = P(E) \cdot P(F)$
Dependent	$P(E \wedge F) = P(F) \cdot P(E|F)$
Mutually Exclusive	$P(E \veebar F) = P(E) + P(F)$
Non-mutually Exclusive	$P(E \vee F) = P(E) + P(F) - P(E \wedge F)$

96. Empirical Probability. In case we cannot determine precisely the number of ways an event can happen or fail to happen in one trial, we shall define a probability based on a statistical record. Suppose that we observe that in a large number of trials, n, an event happened s times. We say that the **relative frequency** of its occurrence is s/n. As n becomes very large the relative frequency changes very little and usually tends to a number. We then define an **empirical probability** that an event will happen in one trial to be s/n. Life insurance rates are based on mortality tables which have been determined from empirical probabilities.

Illustration. Data collected on 2,000 people who entered a drug store revealed that 250 bought cosmetics and 400 bought cigarettes. What is the probability that a person will buy (a) cosmetics, (b) cigarettes, (c) one or the other.

Solution.

a) $p = \dfrac{250}{2000} = \dfrac{1}{8}$

b) $p = \dfrac{400}{2000} = \dfrac{1}{5}$

c) The events are mutually exclusive; therefore

$$p = \frac{1}{8} + \frac{1}{5} = \frac{13}{40}$$

97. Expectation. If p is the probability of success in one trial of an event, then *pk* is the **probable number** of successes in k trials. If a person is to receive a sum of S dollars in case a certain event with a probability p results successfully, then the value of his **expectation** is *pS* dollars.

Illustrations.
1) If the probability of throwing a 7 with a pair of dice is 1/6, what is the probable number in 6 tries?
Solution. The probable number is 6 (1/6) = 1.
2) If throwing a 7 with a pair of dice pays \$3, what is the value of the expectation on a single throw?
Solution. The probability of throwing a 7 is 1/6. Therefore the value of the expectation if \$3 (1/6) = \$0.50.

98. Repeated Trials. Let us consider the question of successive trials of an event and let p be the probability that a specified situation will occur in a given trial. Then $q = 1 - p$ is the probability that the situation will fail to occur in this trial. Suppose that n trials are made and that the situation occurs in *exactly* r trials and fails to occur in the other $n - r$ trials. Since the trials are independent and since r trials can be selected from n trials in $C(n, r)$ ways, the probability that this will happen is $C(n, r)\, p^r q^{n-r}$.

The specified situation will occur *at least* m times if it occurs n, $n - 1$, $n - 2$, ..., $m + 1$, m times. These events are mutually exclusive and the probability that the situation will occur at least m times in n trials is the sum of the exact probabilities with $r = n$, $n - 1$, ..., m,

$$p^n + C(n, n - 1)p^{n-1}q + \ldots + C(n, m)p^m q^{n-m}$$

Illustration. A single die is tossed 4 times. What is the probability that a 2 will appear (a) exactly twice, (b) at least twice, and (c) at least once?

Solution. A die has 6 faces, so the probability of a 2 is $p = 1/6$ and $q = 5/6$ with $n = 4$.

a) Exactly twice; $r = 2$.

$$P = C(4, 2)\left(\frac{1}{6}\right)^2 \left(\frac{5}{6}\right)^2$$

$$= \frac{4 \cdot 3}{2}\left(\frac{1}{36}\right)\left(\frac{25}{36}\right) = \frac{25}{216}$$

b) At least twice; $m = 2$.

$$P = \left(\frac{1}{6}\right)^4 + C(4, 3)\left(\frac{1}{6}\right)^3 \left(\frac{5}{6}\right) + C(4, 2)\left(\frac{1}{6}\right)^2 \left(\frac{5}{6}\right)^2$$

$$= \frac{1}{6^4}\left[1 + 20 + 150\right] = \frac{171}{1296}$$

c) At least once; $m = 1$.

$$P = \frac{1}{6^4}\left[1 + 20 + 150 + 500\right] = \frac{671}{1296}$$

Summary of Symbols

$n(A)$, number of elements in the set A

$P(n, r)$, permutation of n things taken r at a time

$C(n, r)$, combination of n things taken r at a time

$\binom{n}{k}$, binomial coefficient $= C(n, k)$.

---

99. Exercise XI.

1. Given the set $S = \{a, b, c, d, e, f\}$. Find the cross-partitions of the following pairs of partitions:

   a) $A_1 = \{a, b, c, d\}$; $A_2 = \{e, f\}$ and $B_1 = \{a, b, c\}$; $B_2 = \{d, e, f\}$

   b) $A_1 = \{a, b\}$, $A_2 = \{c, d\}$, $A_3 = \{e, f\}$, and $B_1 = \{a, b\}$; $B_2 = \{c, d, e, f\}$

2. Given a set of nine objects, eight of which have the same weight and one is heavier. Show how, in two weighings with a pan balance, the heavy one can be identified.

3. Given $A = \{1, 2, 3, 4, 5\}$ and $B = \{3, 4, 5, 6, 7\}$. Find the number of elements in $A \cup B$.

4. In a group of 50 students, 23 take mathematics, 32 take English, 11 take chemistry, 9 take mathematics and chemistry, 11 take English and mathematics, and 2 take all three. How many of the

50 students do not take any of the three subjects?

5. Draw a Venn diagram and show that

$$n(A \cup B \cup C) = n(A) + n(B) + n(C) - n(A \cap B) - n(A \cap C)$$
$$- n(B \cap C) + n(A \cap B \cap C)$$

6. Find the values of the following permutations:
   a) $P(9, 4)$ b) $(P(11, 3)$ c) $P(30, 4)$

7. How many three-digit numbers can be formed with the digits 1, 2, 3, and 4 if (a) no digit is repeated, (b) repetitions are permitted?

8. In how many ways can the letters of the word *Tennessee* be arranged?

9. The Greek alphabet has 24 letters. How many fraternity names, each of three letters, can be formed if repetitions are permitted?

10. How many odd numbers of two digits can be formed from the digits 1, 2, 3, and 4?

11. Solve for n:
    a) $P(n, 3) = 504$ b) $P(n, 3) = 210$

12. Compute:
    a) $C(12, 5)$ b) $C(9, 3)$ c) $C(n, n - 1)$

13. How many different triangles can be determined by ten points, no three of which lie in the same straight line?

14. There are 24 swimmers entered in the free-style preliminary in a pool with 8 lanes. In how many ways can the contestants for the first heat be selected?

15. A committee of five is to be selected from eight Democrats and seven Republicans. In how many ways can the selection be made if the committee is to contain no more than two Republicans?

16. An ordinary die has 6 faces with numbers from 1 to 6. How many possible combinations can occur with a pair of dice? With 3 dice?

17. What is the probability of rolling, with a pair of dice, (a) 2, (b) 3, (c) 4, (d) 5, (e) 6, (f) 7, (g) 8, (h) 9, (i) 10, (j) 11, (k) 12?

18. Solve for n:
    a) $P(n, 2) = 6C(n, 3)$ b) $6P(n, 2) = 12 C(n, 4)$

19. A bag contains 7 white balls, 9 black balls, 8 red balls, and 12 blue balls. What is the probability that on two consecutive draws both balls will be red if the first ball drawn (a) is replaced, (b) is not replaced?

20. If a pair of dice is tossed twice, what is the probability that
    a) a 5 is obtained on both tosses,

   b) a 6 is obtained on both tosses,
   c) a 7 is obtained on both tosses,
   d) a 6 is obtained on the first toss and a 7 is obtained on the
      second toss?
21. At a busy intersection 20,000 cars were tabulated as follows:
    14,000 regular passenger cars, 2,000 trucks, and 4,000 compact
    cars. What is the probability that the next car is a compact car?
22. In 5 tosses of a pair of dice, what is the probability that a 7 will
    appear (a) exactly 3 times, (b) at least 3 times?
23. If a player gets $10 each time he throws a 7, what is the value of
    the expectation in problem 22(b)?
24. There is one winning ticket in a box containing 100 tickets and
    it pays $50. What is the value of the expectation of a single draw?
25. Consider the bag of Problem 19. What is the probability that in
    7 draws (the ball being replaced each time) a blue ball will be
    drawn at least 3 times?

X

A BIT OF LOGIC

100. Introduction. Throughout this book we have presented definitions, axioms, properties, and theorems in the form of statements. These statements have helped us build a logical structure for elementary algebra. At the beginning of Chapter II (Sections 17 and 18) we discussed the most frequently used types of statements. This discussion was sufficient to develop our algebra. We can, however, study the statements themselves in greater detail.

Consider the statement

It is hot and it is very humid.

This is a compound statement made up of the two parts "It is hot" and "it is very humid." These simple statements are joined by the connective "and." If we know something about the truth of each simple statement, what can we say about the truth of the compound statement? What happens if we change the connective? These are questions which are considered in mathematical logic.

101. Symbols. In the analysis of compound statements, mathematicians make extensive use of symbols. We start with the following:

$p, q, \ldots$ simple statements	$\rightarrow$ implies
$\wedge$ and	$\leftrightarrow$ is equivalent to
$\vee$ or	$\sim$ not

We then define

$p \wedge q$ as the **conjunction** of p and q,
$p \vee q$ as the **disjunction** of p and q,
$p \rightarrow q$ as the **implication**, (p implies q),
$p \leftrightarrow q$ as the **equivalence** relation,
$\sim p$ as the **negation** of p.

204

The disjunction can sometimes be ambiguous. The statement "I will call Bob or Jim" carries the possibility that I will call both men. This is the *inclusive disjunction* (*p* or *q* or both) and is the usual meaning. The *exclusive disjunction* means *p* or *q* but *not* both; this is denoted by $\veebar$.

Examples:
a) 2 and 4 are even numbers; $p \wedge q$.
b) The number *n* is even or odd; $p \veebar q$.
c) If *x* is even, then x^2 is even; $p \rightarrow q$.
d) $x + 3 = 5$ if and only if $x = 2$; $p \leftrightarrow q$.
e) It is hot and it is not raining; $p \wedge \sim q$.

102. Truth Tables. Since the truth of a compound statement depends upon the truth of its components, we can form tables of all the possibilities for each connective. Such tables are called **truth tables**. In constructing them we use T to mean true and F to mean false. The simplest is the negation:

p	$\sim p$
T	F
F	T

This table is formed by listing the possibilities for *p* (T and F) and determining the corresponding truth of the last column. It should be clear that if *p* is true then *not-p* is false and that the reverse also holds. The truth tables for the four simple connectives are on p. 206.* These tables will be used to form other truth tables.

Illustrations. Use the truth tables to establish the truth of the following statements.
1) The numbers 2 and 3 are even numbers.
Solution. $\quad\quad\quad$ *p* = "2 is an even number," T
$\quad\quad\quad\quad\quad$ *q* = "3 is an even number," F
$\quad\quad\quad\quad\quad$ $p \wedge q$ is F by the truth table.

---

*If p is false in the implication $p \rightarrow q$, then the truth of the compound statement cannot be challenged regardless of the truth of q, and the statement is said to be true. Consider the statement, "If you hit me, I shall yell." The statement remains true when you do not hit me whether I yell or not.

p	q	$p \wedge q$
T	T	T
T	F	F
F	T	F
F	F	F

p	q	$p \vee q$
T	T	T
T	F	T
F	T	T
F	F	F

p	q	$p \rightarrow q$
T	T	T
T	F	F
F	T	T
F	F	T

p	q	$p \leftrightarrow q$
T	T	T
T	F	F
F	T	F
F	F	T

2) One of the numbers 2 or 3 is an even number.
Solution. $p =$ "2 is even," T
                                     $q =$ "3 is even," F
                          $p \vee q$ is T by the truth table.

3) If n is an integer, then $2n$ is an even integer.
Solution. $p =$ "n is an integer"
                          $q =$ "$2n$ is an even integer"

p	q	$p \rightarrow q$
T	T	T

4) $\sim (\sim p) \leftrightarrow p$
Solution.

p	$\sim p$	$\sim (\sim p)$	$\sim (\sim p) \leftrightarrow p$
T	F	T	T
F	T	F	T

Truth tables may be used to establish the basic laws of logic. We need the following definition.

Definition. *A* **tautology** *is a* **true** *proposition formed by combining other propositions* p, q, r, . . . *regardless of the truth of these components.*

Illustration 4 above is an example of a tautology and is the Double Negation Law. The proof is accomplished by a truth table with its first columns representing propositions $p, q, r, \ldots$, additional columns representing parts of the proposition, and a last column representing the complete proposition. The number of rows depends upon the number of propositions $p, q, r, \ldots$; since we have two for each (T and F), the number of rows is 2^n where n is the number of arbitrary propositions. If *every* entry in the last column is T, the proposition is a tautology.

Theorem. *Law of Excluded Middle.* $p \vee (\sim p)$
Proof. (For the last column see truth table for. $\vee$.)

p	$\sim p$	$p \vee (\sim p)$
T	F	T
F	T	T

Theorem. *Law of Contradiction. A proposition and its negation are not both true; i.e.,* $\sim [p \wedge (\sim p)]$ *is a tautology.*
Proof.

p	$\sim p$	$p \wedge (\sim p)$	$\sim [p \wedge (\sim p)]$
T	F	F	T
F	T	F	T

Theorem. *Law of Syllogism. If* p *implies* q *and* q *implies* r, *then* p *implies* r; $[(p \to r) \wedge (q \to r)] \to (p \to r)$
Proof.

(1)	(2)	(3)	(4)	(5)	(6)	(7)	(8)
p	q	r	$p \to q$	$q \to r$	(4) $\wedge$ (5)	$p \to r$	(6) $\to$ (7)
T	T	T	T	T	T	T	T
T	T	F	T	F	F	F	T
T	F	T	F	T	F	T	T
T	F	F	F	T	F	F	T
F	T	T	T	T	T	T	T
F	T	F	T	F	F	T	T
F	F	T	T	T	T	T	T
F	F	F	T	T	T	T	T

103. Converse, Inverse, and Contrapositive. We have used these three concepts throughout the book. Let us consider them from the viewpoint of this chapter. Given the implication $p \to q$, then

Converse: $q \to p$

Inverse: $(\sim p) \to (\sim q)$

Contrapositive: $(\sim q) \to (\sim p)$

Theorem. *The converse of a true implication is not always true.*
Proof.

p	q	$p \to q$	$q \to p$
T	T	T	T
T	F	F	T
F	T	T	F
F	F	T	T

Note that in the third row $p \to q$ is true but $q \to p$ is false.

Theorem. *The inverse of a true implication is not always true.*
Proof.

p	q	$\sim p$	$\sim q$	$p \to q$	$(\sim p) \to (\sim q)$
T	T	F	F	T	T
T	F	F	T	F	T
F	T	T	F	T	F
F	F	T	T	T	T

Theorem. *An implication and its contrapositive are equivalent.*
Proof.

p	q	$\sim p$	$\sim q$	$p \to q$	$(\sim q) \to (\sim p)$
T	T	F	F	T	T
T	F	F	T	F	F
F	T	T	F	T	T
F	F	T	T	T	T

The last two columns are identical; therefore the two implications are simultaneously true or false.

> **Illustration.** Test the converse, inverse, and contrapositive of the implication, "If two triangles are congruent, then they are similar."
> *Solution.*
> Converse: If two triangles are similar, then they are congruent; F.
> Inverse: If two triangles are not congruent, then they are not similar; F.
> Contrapositive: If two triangles are not similar, then they are not congruent; T.

Note that the inverse $(\sim p) \to (\sim q)$ is the contrapositive of the converse $q \to p$. Consequently, the converse and the inverse are equivalent, i.e., both are true or false together.

104. Theorems and Proofs. Throughout this book we have stated theorems and proofs. In this section we shall present the logic of these statements and formalize our methods. Practically all theorems are stated in the form of an implication or an equivalence. We can therefore use some of the basic logic developed in the last sections.

In the development of proofs we shall employ three fundamental tools, stated as rules of operations.

Rule 1. *Substitution for Variables.* If a proposition p, contains a variable x and is true for *all* x in a given set A, then x may be replaced by any specific member $x_1 \in A$ and p is still true.

> **Illustration.** If $x + 5 = 5 + x$ for all $x \in R$, show that it is true for $x_1 \in R$.
> *Solution.* Let $x_1 = 2$ and replace x by x_1 to obtain $2 + 5 = 7 = 5 + 2$.

This rule can be extended to more than one variable.

> **Illustration.** If $x^2 - y^2 = (x + y\,(x - y)$ for all x and y in the real-number system, show that it is true for $x = 2$ and $y = 3$.
> *Solution.*
> $$(2)^2 - (3)^2 = (2 + 3)(2 - 3)$$
> $$4 - 9 = 5(-1)$$
> $$-5 = -5$$

Although we need not take $x_1 = y_1$ in the above illustration, we cannot let their values change during the substitution.

Rule 2. *Substitution for Propositions.* Any proposition may be replaced by an equivalent proposition.

> **Illustration.** Restate the proposition
> $$(p \to q) \lor (q \to r)$$
> *Solution.* Since an implication and its contrapositive are equivalent, we have
> $$(p \to q) \leftrightarrow [(\sim q) \to (\sim p)]$$
> By substitution, the given proposition can be written in the form
> $$[(\sim q) \to (\sim p)] \lor (q \to r)$$

Let us consider an illustration which may be more familiar.

> **Illustration.** The area of a circle is pi times the square of the radius. Restate this proposition.
> *Solution.* Since the radius is equal to one-half the diameter, we can make the substitution and write
> $$A = \frac{1}{4} \pi D^2$$

Rule 3. *Detachment.* The rule of detachment states that if p is true and if $p \to q$ is true, then we assert that q is true. This rule can be demonstrated by considering the basic truth table.

p	q	$p \rightarrow q$
T	T	T
T	F	F
F	T	T
F	F	T

We are given that p is true and the implication is true; these two facts coincide in the first row only and in this row q is true.

Illustration. Given: If $2k + 1$ is an odd number, then $2k + 2$ is an even number; and, if $2k + 2$ is an even number, then $2k + 3$ is an odd number. Prove the conclusion that if $2k + 1$ is an odd number, then $2k + 3$ is an odd number.

Solution. In our symbolic language

$p =$ "$2k + 1$ is an odd number,"
$q =$ "$2k + 2$ is an even number,"
$r =$ "$2k + 3$ is an odd number."

The theorem states

$$[(p \rightarrow q) \wedge (q \rightarrow r)] \rightarrow (p \rightarrow r)$$

The hypothesis states that $p \rightarrow q$ is true and that $q \rightarrow r$ is true; hence the conjunction $(p \rightarrow q) \wedge (q \rightarrow r)$ is true (see the truth table for conjunctions). The theorem is a syllogism and we have shown that the Law of Syllogism is a tautology and thus is always true. Therefore, by detachment we assert that $p \rightarrow r$ is true, proving the conclusion.

There is usually more than one way to prove or disprove a theorem. The various methods can be gathered into specific categories. (Mathematicians interested in mathematics for its own sake rather than for its applications are continuously searching for the "elegant" proof.)

I. *Direct method.* This is the method which finds greatest application. The procedure is to start with what is given in the hypothesis and apply the rules discussed above until we arrive at the conclusion.

Illustration. Prove that if p and q are true, then $(p \wedge q) \rightarrow (p \vee q)$ is true.

Solution. Since p and q are true, we have $p \wedge q$ and $p \vee q$ are true by the truth table. Substituting these propositions into the implication, we have the first line of the truth table for an implication; thus $(p \wedge q) \rightarrow (p \vee q)$ is true.

II. *Indirect method.* An indirect proof begins by assuming that the given proposition is false. This is the same as assuming that its negation is true. The direct method is then applied to the negation until we arrive at a contradiction which is dependent upon this assumption. On the basis of this contradiction we reason that our assumption—that the proposition is false—is incorrect and, consequently, that the given proposition must be true. Since this method uses the negation of a statement, let us summarize the negations of the four basic statements.

Connective	Symbolism	Negation
Conjunction	$p \wedge q$	$(\sim p) \vee (\sim q)$
Disjunction	$p \vee q$	$(\sim p) \wedge (\sim q)$
Implication	$p \rightarrow q$	$p \wedge (\sim q)$
Equivalence	$p \leftrightarrow q$	$p \leftrightarrow (\sim q)$ or $(\sim p) \leftrightarrow q$

Illustration.   Prove that if p and q are true, then $(p \wedge q) \rightarrow (p \vee q)$ is true.

Solution.   Suppose the implication is false; then the negation $\sim [(p \wedge q) \rightarrow (p \vee q)]$ is true. This negation states that

$$(p \wedge q) \wedge \sim (p \vee q) \qquad \text{or} \qquad (p \wedge q) \wedge (\sim p \wedge \sim q)$$

is true. However, since p and q are true, we have

p	q	$p \wedge q$	$\sim p$	$\sim q$	$\sim p \wedge \sim q$
T	T	T	F	F	F

and $(p \wedge q) \wedge (\sim p \wedge \sim q)$ is false. Thus we have a contradiction, which we shall blame on our assumption that the given implication is false. We consequently assert that the given implication is true.

III. *Proof by enumeration.* If a theorem has only a finite number of possibilities, it can be proved by examining every possibility.

Illustration.   The square of an odd number is an odd number.

Proof.   Every odd number can be expressed in the form $2n + 1$ where

$n = 0, 1, 2, \ldots$. We examine the square of this number.

$$(2n + 1)^2 = 4n^2 + 4n + 1 = 4(n^2 + n) + 1$$

Divide by 2 to obtain $\qquad 2(n^2 + n) + \dfrac{1}{2}$

Since the square is not exactly divisible by 2, it is an odd number.

The above illustration is very simple since we needed to investigate only one case.

Illustration.   If the square of an integer is divided by 5, the remainder is never 3.

Proof.   All of the integers are included in one of the following forms

$$5k - 2, 5k - 1, 5k, 5k + 1, 5k + 2$$

The squares are

$$(5k - 2)^2 = 25k^2 - 20k + 4$$
$$(5k - 1)^2 = 25k^2 - 10k + 1$$
$$(5k)^2 = 25k^2$$
$$(5k + 1)^2 = 25k^2 + 10k + 1$$
$$(5k - 2)^2 = 25k^2 + 20k + 4$$

When these squares are divided by 5, the remainders are 4, 1, and 0. (Note that we could state a stronger theorem specifying that the remainders are only these three numbers.)

The truth tables are good examples of proof by enumeration.

IV. *Disproof by counterexample.* A proposition may be false. This becomes obvious if the negation is proved to be true. Another way to prove a proposition false is to find a counterexample.

Illustrations.

1) Show that the proposition $x + 3y = 7x - 3$ for all x and y is false.

Solution.   Since the equation can be written in the form $y = 2x - 1$ its graph is a straight line; therefore an infinite number of pairs (x, y) satisfies the equation. However, the pair $(1, 2)$ yields

$$2 = 2(1) - 1 = 1$$

which is false. Hence, the equation is not true for all x and y.

2) Show that the proposition "If ab is even, then both a and b are even" is false.

Solution.   Let $ab = 6$, which is even; then $a = 2$ and $b = 3$. Since b is not even, the proposition is false.

We have presented the general methods of proof. There are, of course, many special techniques using these rules and procedures.

Mathematical induction (see Section 36) is such a special technique.

If the proposition uses the quantifier "some" instead of "all" we need only exhibit one element which makes the proposition true.

Illustration. There is a real number which satisfies the equation $3x = 5$.
Solution. To prove this proposition we need only exhibit an $x_1 \in R$. Since 3 and 5 are real numbers and since the quotient of two real numbers is real, we have $x_1 = 5/3 \in R$, which satisfies the equation and proves the proposition.

$$3\left(\frac{5}{3}\right) = 5 = 5$$

105. A Two-Valued Algebra.* Let us return to the concept of truth tables and let "true" be represented by 1 and "false" by 0. Let us further define two operations, which we shall call *addition* and *multiplication*, by the two tables

<table>
<tr><td colspan="3">Addition</td><td colspan="3">Multiplication</td></tr>
<tr><td></td><td>0</td><td>1</td><td></td><td>0</td><td>1</td></tr>
<tr><td>0</td><td>0</td><td>1</td><td>0</td><td>0</td><td>0</td></tr>
<tr><td>1</td><td>1</td><td>1</td><td>1</td><td>0</td><td>1</td></tr>
</table>

In other words, $0 + 0 = 0$ and $0 + 1 = 1 + 0 = 1 + 1 = 1$, and $(0)(0) = (0)(1) = (1)(0) = 0$ and $(1)(1) = 1$. We are considering a set of two elements with two operations. They satisfy the following axioms of real numbers:

Axiom R_1. The *closure* axioms are satisfied since for any combination of 0 and 1 the sum and product are members of the set.

Axiom R_2. The *commutative* axioms are satisfied. This can be shown by enumeration.

Axiom R_3. The *associative* axioms are satisfied. This also can be shown by enumeration.

Axiom R_4. The *distributive* axioms are satisfied. Again this can be shown by enumeration.

Axiom R_5. The *identity* axioms are satisfied. 0 is the identity for addition and 1 is the identity for multiplication.

Our operations also satisfy the *Idempotent Law* which states that for every element of the set the operation of a by a yields a; i.e.,

---

*This algebra was developed largely by the English mathematician George Boole and is often referred to as Boolean algebra.

$0 + 0 = 0$ and $(0)(0) = 0$ as well as $1 + 1 = 1$ and $(1)(1) = 1$.

Since the set contains only two elements the complement of one must be the other; i.e., $0' = 1$ and $1' = 0$. In this two-valued algebra practically all theorems can be proved by enumeration since we have only two elements to consider.

DeMorgan's Theorems.   *The complement of a sum of two elements is the product of their complements.* $(a + b)' = a'b'$

The complement of a product of two elements is the sum of their complements. $(ab)' = a' + b'$

Proof. We shall enumerate the possibilities and show that the columns headed by the quantities on each side of the equal sign are identical. This is the same procedure we used with truth tables.

a	b	a'	b'	$(a + b)$	$(a + b)'$	$a'b'$	ab	$(ab)'$	$a' + b'$
0	0	1	1	0	1	1	0	1	1
0	1	1	0	1	0	0	0	1	1
1	0	0	1	1	0	0	0	1	1
1	1	0	0	1	0	0	1	0	0

Illustration.   Prove that for all a, b, and c, we have

$$a + bc = (a + b)(a + c).$$

Solution.

a	b	c	bc	$a + bc$	$a + b$	$a + c$	$(a + b)(a + c)$
0	0	0	0	0	0	0	0
0	0	1	0	0	0	1	0
0	1	0	0	0	1	0	0
0	1	1	1	1	1	1	1
1	0	0	0	1	1	1	1
1	0	1	0	1	1	1	1
1	1	0	0	1	1	1	1
1	1	1	1	1	1	1	1

106. Veitch Diagrams.   Proof by enumeration can become very tedious if a large number of cases has to be considered. Diagrams can be useful in such instances and we shall introduce the Veitch diagram, which is used in the algebra of sets.

Let us consider a universe U with two subsets A and B. Divide

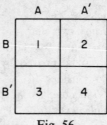

Fig. 56

a square representing U into four equal parts (see Figure 56). Let the *left* half represent A and let the *upper* half represent B. Number the four subsquares 1, 2, 3, and 4, as shown in Figure 56. The elements in 1 are in both A and B and therefore are in the intersection $A \cap B$, which we shall also denote as the product AB. The elements in 2 are in B but not in A; we can denote this subset by BA'. The elements in 3 are in A but not in B; hence, they are in AB'. The elements in 4 are neither in A nor in B, so they are in $A'B'$. Thus every point in U is in one of these subsquares. By placing an x in the proper subsquare we can indicate a particular situation. The empty set, $\emptyset$, will have no x anywhere and the set U will have an x in all four subsquares. Figure 57 shows the four sets A, A', B, and B'. Figure 58 shows the four products and Figure 59 shows the four sums.

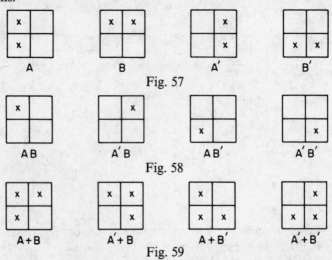

Properties can now be verified by considering the corresponding Veitch diagrams. Consider, for example, DeMorgan's theorems.

Since [diagram: A+B], then [diagram: (A+B)'], which is the same as the diagram

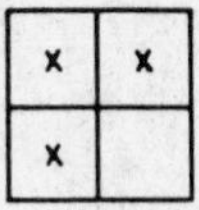

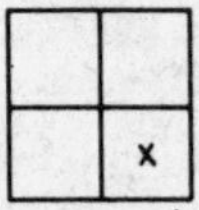

for $A'B'$ (Figure 58).

Illustration. Verify that $AB' + B = (A'B')' = A + B$.

Solution. The Veitch diagrams are

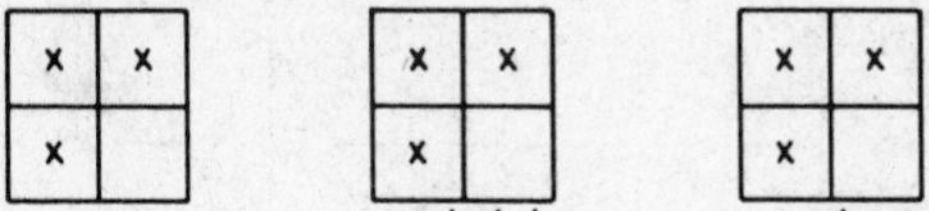

[diagrams]

A + B (A'B')' A B' + B

Fig. 60

The extension to three subsets A, B, C is accomplished by letting A be the left half, B be the upper half, and C be the middle half. See Figure 61.

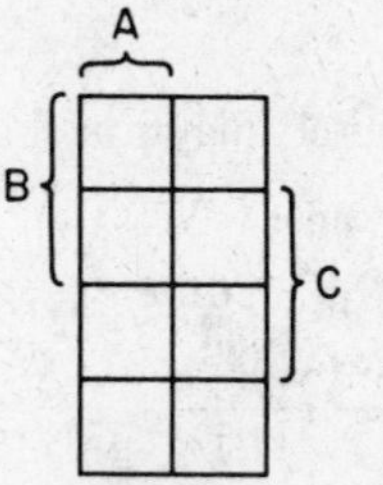

Illustration. Verify that
$$A + BC = (A + B)(A + C).$$

Solution.

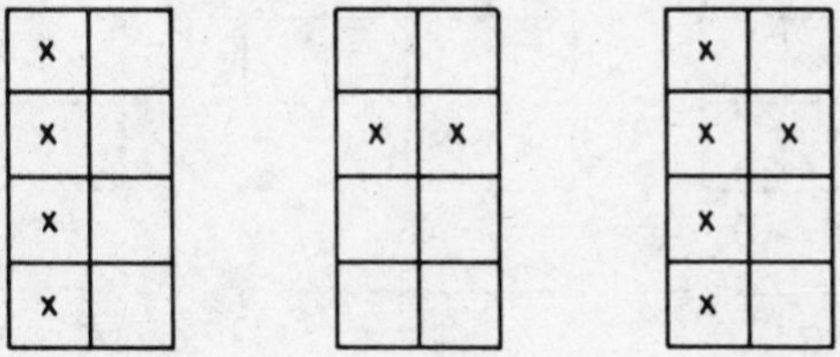

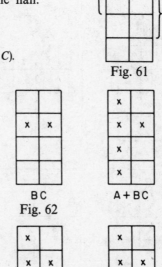

Fig. 61

[diagrams]

A   BC   A + BC

Fig. 62

[diagrams]

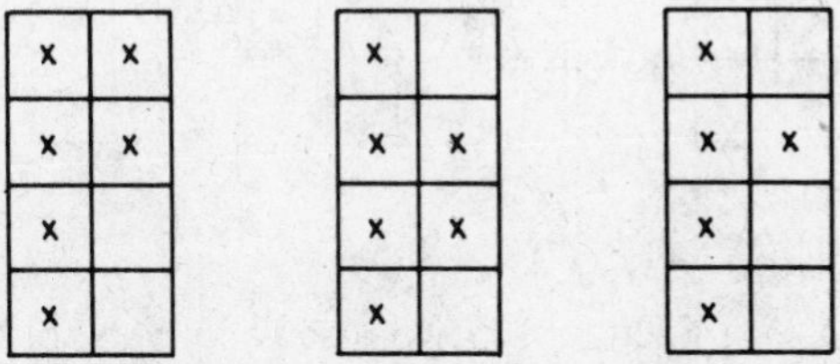

A + B   A + C   (A + B)(A + C)

Fig. 63

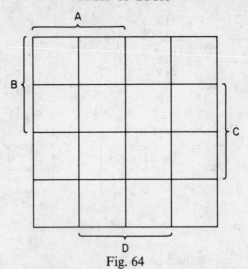

Fig. 64

The Veitch diagram for four subsets is shown in Figure 64.

Illustration. Verify that

$$(A + B)(A + C) + BD = A + B(A + C + D)$$

Solution.

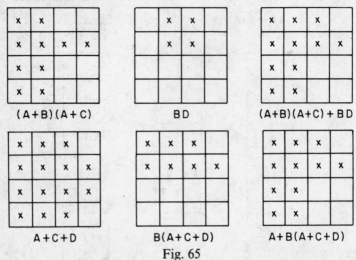

Fig. 65

107. Groups. In Section 105 we discussed, very briefly, the idea of a two-valued algebra. In developing the idea we checked the axioms of the real-number system. This is a usual procedure in the development of a logical structure. We can in fact build an abstract mathematical system based on a specified set of axioms. The axioms will involve a set of elements, $a, b, c, \ldots \in G$, and an operation to pair two elements, $a \circ b$. We have already considered the numbers as elements and the four operations of arithmetic in the development of the algebra presented in this book. Let us consider a more abstract situation.

Definitions. *A mathematical structure which satisfies the closure, associative, identity, and inverse axioms is called a* **group.** *If the system also satisfies the commutative axiom, it is called a* **commutative group.**

Since the real numbers satisfy these axioms with respect to addition, they form a commutative group. If we exclude the number 0, they also form a commutative group with respect to multiplication.

Illustration. Show that the set $G = \{1, 2\}$ forms a commutative group with respect to multiplication defined by the table

x	1	2
1	1	2
2	2	1

Solution. Test the axioms.

1. *Closure.*
$$1 \times 1 = 1 \qquad 1 \times 2 = 2$$
$$2 \times 1 = 2 \qquad 2 \times 2 = 1$$

Since these are all in G the closure axiom is satisfied.

2. *Associative.*
$$(1 \times 1) \times 1 = 1 \times (1 \times 1) = 1$$
$$(1 \times 1) \times 2 = 1 \times (1 \times 2) = 2$$
$$(1 \times 2) \times 1 = 1 \times (2 \times 1) = 2$$
$$(1 \times 2) \times 2 = 1 \times (2 \times 2) = 1$$
$$(2 \times 1) \times 1 = 2 \times (1 \times 1) = 2$$
$$(2 \times 1) \times 2 = 2 \times (1 \times 2) = 1$$
$$(2 \times 2) \times 1 = 2 \times (2 \times 1) = 1$$
$$(2 \times 2) \times 2 = 2 \times (2 \times 2) = 2$$

3. *Identity.* The identity element is 1 since $1 \times 1 = 1 \times 1 = 1$ and $1 \times 2 = 2 \times 1 = 2$.

4. *Inverse.* The inverse of a is a since $1 \times 1 = 1$ and $2 \times 2 = 1$.

5. *Commutative.*
$$1 \times 1 = 1 = 1 \times 1$$
$$1 \times 2 = 2 = 2 \times 1$$
$$2 \times 2 = 1 = 2 \times 2$$

There is an endless number of examples of a group under the familiar operations of addition and multiplication. Let us consider an example with a different type of operation. Let the set G contain three elements a, b, and c and consider the elements to be ordered in that relative position. There are six possible permutations of the three elements, as follows,

$$e = \begin{pmatrix} a & b & c \\ a & b & c \end{pmatrix} \qquad p = \begin{pmatrix} a & b & c \\ c & a & b \end{pmatrix} \qquad q = \begin{pmatrix} a & b & c \\ b & c & a \end{pmatrix}$$

$$r = \begin{pmatrix} a & b & c \\ a & c & b \end{pmatrix} \qquad s = \begin{pmatrix} a & b & c \\ b & a & c \end{pmatrix} \qquad t = \begin{pmatrix} a & b & c \\ c & b & a \end{pmatrix}$$

which show the rearrangement from the original order (a, b, c). Thus p replaces a by c, b by a, and c by b, etc. The operation $p \circ r$ specifies that we first carry out the permutation p and then the permutation r.

$$p \circ r = \begin{pmatrix} a & b & c \\ c & a & b \end{pmatrix} \circ \begin{pmatrix} a & b & c \\ a & c & b \end{pmatrix} = \begin{pmatrix} a & b & c \\ b & a & c \end{pmatrix} = s$$

We can now show that the set $G = \{e, p, q, r, s, t\}$ forms a group under this operation. The closure and associative axioms can be established by enumeration. The identify element is e. The inverse for p is q since

$$p \circ q = \begin{pmatrix} a & b & c \\ c & a & b \end{pmatrix} \circ \begin{pmatrix} a & b & c \\ b & c & a \end{pmatrix} = \begin{pmatrix} a & b & c \\ a & b & c \end{pmatrix} = e$$

The inverse for q is p, and r, s, and t are their own inverses. The group is not commutative for

$$r \circ p = \begin{pmatrix} a & b & c \\ a & c & b \end{pmatrix} \circ \begin{pmatrix} a & b & c \\ c & a & b \end{pmatrix} = \begin{pmatrix} a & b & c \\ c & b & a \end{pmatrix} = t$$

and we have already seen that $p \circ r = s$.

108. Fields. In the last section we discussed a mathematical system based on one operation "$\circ$". We shall now consider one with two operations and use the familiar "$+$" and "$\times$" to denote the operations.

Definition. *A **field** is a mathematical system consisting of a set of objects, F, and two binary operations "$+$" and "$\times$" and satisfying the closure, associative, identity, inverse, and commutative axioms for each operation and the distributive law for joining the two operations.*

Thus a field is an abstract mathematical system consisting of a set of elements, two operations, and eleven axioms. Symbolically we can write for $F(a, b, c, e, E, \ldots)$

Operation +	Name	Operation $\times$
$\mathbf{R_1}.\ a + b = c$	Closure	$\mathbf{R_6}.\ a \times b = c$
$\mathbf{R_2}.\ (a + b) + c =$ $\quad a + (b + c)$	Associative	$\mathbf{R_7}.\ (a \times b) \times c =$ $\quad a \times (b \times c)$
$\mathbf{R_3}.\ a + e = e + a = a$	Identity	$\mathbf{R_8}.\ a \times E = E \times a$ $\quad = a$
$\mathbf{R_4}.\ a + (-a) =$ $\quad (-a) + a = e$	Inverse	$\mathbf{R_9}.\ a \times \dfrac{1}{a} = \dfrac{1}{a} \times a = E,$ $\quad a \neq e$
$\mathbf{R_5}.\ a + b = b + a$	Commutative	$\mathbf{R_{10}}.\ a \times b = b \times a$
Distributative Law $\mathbf{R_{11}}.\ a \times (b + c) = (a \times b) + (a \times c)$		

Since the rational numbers, the real numbers, and complex numbers satisfy the axioms of the definition, they each form a field. A great part of algebra is the study of the properties of a field. To show that a given system forms a field requires the verification of all eleven axioms. Although this may be time consuming it does establish at the same time all the generalized theorems of a field for the particular system under consideration. For example, once we have proved the theorem "a linear equation, $ax + b = 0$, $a \neq 0$, has a unique solution in any field," then given a field we know that a linear equation in this field has a unique solution. The subject of algebra could be completely developed in this abstract manner.

There are many examples of fields, some with an infinite number of elements, others with only a finite number.

Summary of Symbols	
$\wedge$, and	$\vee$, or (inclusive)
$\rightarrow$, implies	$\leftrightarrow$, equivalent
$\sim$, not	$\veebar$, exclusive or
T, true	F, false
$\circ$, operation	A', complement of A.

109. Exercise XII.

1. Construct a truth table for each of the following:

   a) $\sim (p \wedge q)$ b) $\sim (p \vee q)$

   c) $p \vee \sim p$ d) $\sim p \wedge \sim q$

   e) $(p \vee q) \vee \sim p$ f) $\sim (p \vee q) \wedge p$

   g) $(p \vee q) \leftrightarrow (q \vee p)$ h) $(p \vee r) \wedge (p \rightarrow q)$

   i) $(p \rightarrow p) \vee (p \rightarrow \sim p)$ j) $(p \rightarrow q) \wedge (q \rightarrow p)$

   k) $(p \rightarrow q) \leftrightarrow \sim (p \wedge \sim q)$ l) $p \wedge q \rightarrow p$

   m) $(p \wedge \sim p) \rightarrow q$ n) $\sim p \vee \sim q$

   o) $p \rightarrow (q \vee r)$ p) $(p \vee \sim q) \wedge r$

2. Which of the following are tautologies?

   a) $(p \vee q) \leftrightarrow \sim (\sim p \wedge \sim q)$ b) $(p \wedge \sim p) \rightarrow q$

   c) $p \vee (q \vee r) \rightarrow (p \vee q) \vee r$ d) $(p \wedge q) \rightarrow (p \vee q)$

   e) $(p \rightarrow q) \leftrightarrow (\sim p \vee q)$ f) $p \wedge (q \wedge r) \leftrightarrow (p \wedge q) \wedge r$

   g) $q \rightarrow (p \vee q)$ h) $[p \rightarrow (q \rightarrow r)] \leftrightarrow (p \rightarrow r)$

3. State the converse, inverse, and contrapositive of the following implications:

   a) If two triangles are congruent, then they are similar.

   b) If a is an even integer, then $2a$ is divisible by 4.

   c) If $2k + 1$ is an odd integer, then $2k + 3$ is an odd integer.

   d) $q \rightarrow p$.

   e) $\sim q \rightarrow \sim p$

   f) $q \rightarrow \sim p$

4. Show that the converse and inverse of a given implication are equivalent.

5. Prove that the sum of two odd integers is even.

6. Find a set of numbers for which
$$(y + x)(y + z) = 2y + xz$$
Prove that the equation is not true for all values of x, y, and z.

7. Prove that if the square of any integer is divided by 3, the remainder cannot be 2.

8. Prove that for our two-valued algebra, the following equations are identities:

   a) $x + xy = x$ b) $x(x + y) = x$

   c) $x(yz) = (xy)z$ d) $x(y + z) = xy + xz$

   e) $xy + xy' + x'y + x'y' = 1$

   f) $(x + y)(x' + y)(x + y')(x' + y') = 0$

9. Use Veitch diagrams to prove the following:

   a) $(AB)' = A' + B' = AB' + A'$

   b) $(A'B)' = A + B' = A + A'B'$

   c) $(AB')' = A' + B = AB + A'$

d) $(A'B')' = A + B = AB' + B$

10. Given the set $G = (0, 1, 2)$ and the addition table

+	0	1	2
0	0	1	2
1	1	2	0
2	2	0	1

show that these elements and operations form a commutative group.

11. The set $G = \{e, a, b, c\}$ with the operation table

∘	e	a	b	c
e	e	a	b	c
a	a	e	c	b
b	b	c	e	a
c	c	b	a	e

is a commutative group called the Klein "Four Group."
 a) Verify that $(a \circ b) \circ c = a \circ (b \circ c)$.
 b) Verify that $a \circ b = b \circ a$.
 c) What are the inverses?

12. Show that $G = \{1, w_1 = -\dfrac{1}{2} + \dfrac{\sqrt{3}}{2} i, w_2 = -\dfrac{1}{2} - \dfrac{\sqrt{3}}{2} i\}$

where $i^2 = -1$ is a commutative group under normal multiplication.

13. Show that our two-valued algebra does not form a field. (Hint: Check for the additive inverse.)

14. Show that the set of real numbers of the form $a + b\sqrt{2}$ where a and b are rational numbers forms a field.

15. Use mathematical induction to prove each of the following:
 a) $\dfrac{1}{1\cdot2} + \dfrac{1}{2\cdot3} + \dfrac{1}{3\cdot4} + \ldots + \dfrac{1}{n(n+1)} = \dfrac{n}{n+1}$
 b) The binomial theorem if n is a positive integer.

EXAMINATIONS

In the pursuit of knowledge one's progress is often measured by examinations. The usual procedure is to consider a set of questions and without any outside aid (except possibly a set of tables) to attempt to arrive at a set of answers. The following examinations have been prepared to aid the reader in assessing his particular state of achievement. These examinations should also prepare a reader to tackle other formal examinations. Short tests on each chapter are followed by examinations on particular topics and on the book as a whole.

Part I. One-Hour Tests

Test #1. (Covers material in Chapter 1.)

1. Peform the indicated operations and simplify:
   a) $\{3(7 - 5) + 2\} \div \{2[4(3 - 2) - 6(5 - 4)] + 2\}$
   b) $(3ax^2y^3)(- 2a^2x)^2 \div 3axy$
   c) $(3x^2 + 5xy - 2y^2) \div (x + 2y)$
   d) $64a^2(xy^2)^3 \div 24(ax)^2y^5$
   e) $(27x^0y^3 - 8x^3y^0) \div (3y - 2x)$
   f) $(\sqrt{x} + 2\sqrt{y})(2\sqrt{x} - \sqrt{y})$
   g) $(3 + 2i) \div (3 - 2i)$
   h) $\dfrac{x + 1}{x^2 - x - 6} \div \dfrac{x + 1}{x + 2}$
   i) $(2 - i)(3 + \sqrt{- 1}) \div (2i + 1)$
   j) $\dfrac{3 - x}{2x^2 + x - 6} + \dfrac{x + 2}{2x^2 - 5x + 3}$

2. Solve the following equations:
   a) $\dfrac{1}{2}(3x - 2) + \dfrac{1}{3}(5x - 1) = 2x$
   b) $\dfrac{1}{2}x + \dfrac{1}{3}x = \dfrac{1}{4}x + \dfrac{1}{5}x + \dfrac{1}{6}x$
   c) $\dfrac{3 + 4x}{4 - 3x} = \dfrac{5 - 8x}{6x + 5}$
   d) $ax - 3x = 7b$
   e) $3x - 17 = 17 - 14x$

3. The sum of two numbers is 38. If the larger number is divided by the smaller, the quotient is 11 and the remainder is 2. Find the two numbers.

Test #2. (Covers material in Chapter 2.)

1. Find the set resulting from the indicated operations:
   a) $[1, 3] \cup [2, 7]$
   b) $\{1, 3, 5, 7\} \cup \{2, 4, 6\}$
   c) $[1, 5] \cap [2, 6]$
   d) $\{1, 2, 3, 5\} \cap \{2, 3, 4\}$
   e) $A \cap A'$
   f) $A \cup A'$

2. Draw Venn diagrams to verify the following:
   a) $A \cap (B \cup B') = A$
   b) If $C' \subset B$, then $B' \subset C$.

3. If $U = \{x | x$ is an even integer $< 10\}$, find A' in each of the following cases:
   a) $A = \{x | x$ is an even integer $< 6\}$
   b) $A = \{x | 2 < x < 4\}$
   c) $A = \{x | x - 3 = 3\}$
   d) $A = \{x | x$ is larger than 1 but less than 5$\}$
   e) $A = \{2, 8\}$

4. Find $A \times B$ if
   a) $A = \{1, 3\}$ and $B = \{2, 4\}$
   b) $A = \{x, y\}$ and $B = \{a, b\}$
   c) $A = \{1, 2, 3\}$ and $B = \{x, y\}$

5. Show that there is a one-to-one correspondence between A and B if A is the set of positive integers and B is the set of negative integers.

Test #3. (Covers material in Chapter 3.)

1. Find the additive and multiplicative inverses of $1, 3, -6, \sqrt{2}$, and $-1/2$.
2. Prove that the set of real numbers is dense.
3. Find the rational number represented by $1.\overline{6}$.
4. Find the set of real numbers given by $]3, 7] \cap [2, 6] \cup [1, 5[$.
5. Perform the following operations:
   a) $(3 + 5i) - (1 - 2i)$
   b) $(3 + i)(i - 3)$
   c) $(7 + 2i) \div (2 - 2i)$

Test #4. (Covers material in Chapter 4.)

1. Given $f(x) = x^2 - 3$ and $g(x) = x - 1$, find the following values at $x = -1, 0, 1, 3$:
   a) $f(x)$
   b) $g(x)$
   c) $f + g$
   d) $f - g$
   e) fg
   f) f/g
   g) $f(g)$

2. Draw the graph of $y = |x| + |-x|$ for $-3 \le x \le 3$.
3. Let f be defined by $y = mx + b$. Find f^{-1} if
   a) $m = 3, b = -2$
   b) $m = -2, b = 3$
   c) $m = 2, b = 0$
   d) $m = b = 1, x < 0$
4. Find $[f(x + h) - f(x)] \div h$ if $f(x) = 3x - 5$.

Test #5. (Covers material in Chapter 5.)

1. Find the slope and intercepts of the straight line defined by $3y - 2x = 12$.

2. Find the slope and equation of the line through the points $P_1(-2, -3)$ and $P_2(3, -1)$.

3. Find the solution set of the system

$$\begin{cases} 3x - 2y - 3 = 0 \\ 3y - 2x + 7 = 0 \end{cases}$$

4. Find the matrix $2AB$ if

$$A = \begin{pmatrix} 1 & -1 \\ 3 & 1 \\ 2 & 5 \end{pmatrix} \quad \text{and} \quad B = \begin{pmatrix} 1 & 2 \\ 2 & 1 \end{pmatrix}$$

5. Find the set of real numbers which satisfies
   a) the linear inequality $12 - 3x \le 0$
   b) the system

$$\begin{cases} x - 2y + 6 \ge 0 \\ y \ge 0 \\ x + 2y - 6 \le 0 \end{cases}$$

Test #6. (Covers material in Chapter 6.)
1. Find the solution set of each of the following equations:
   a) $3x^2 - 6x - 1 = 0$
   b) $2x^2 + 3x + 4 = 0$

2. Sketch the graph of each of the following equations and find the common solution set:
$$2y^2 = x \quad \text{and} \quad x - y^2 = 1$$

3. Find the inverse of $f: x^2 + y = 4$ if $d_f: 0 \le x \le 4$.
4. Find the solution set of $x^3 = 1$.
5. Find the solution set of $x^2 + x - 6 < 0$.
6. Resolve

$$\frac{x^2 + 1}{x^2 - 2x + 1}$$

   into the simplest partial fractions.
7. Sketch the function

$$y = \frac{4}{x^2 - 4}$$

Test #7. (Covers material in Chapter 7.)
1. Express as the algebraic sum of logarithms

$$\log \frac{g\sqrt{h^3}}{t^2 v}$$

2. Given $\log 7 = 0.84510$ and $\log 3 = 0.47712$, find $\log_3 7$.

3. Find the inverse of $y = \sinh x$.
4. Find the domain for which $\log (3x - 5)$ is positive.
5. Prove the identity $\cosh^2 x - \sinh^2 x = 1$.

Test #8. (Covers material in Chapter 8.)
1. Insert six arithmetic means between 3 and 8.
2. Insert four geometric means between 2 and 64.
3. Find the first four terms of $(2x - 3)^{\frac{1}{2}}$.
4. Find the interval of convergence for
$$1 - \frac{x^2}{2!} + \frac{x^4}{4!} - \frac{x^6}{6!} + \cdots$$
5. Use the series for e^x and e^{-x} to derive the series for $\cosh x$.

Test #9. (Covers material in Chapter 9.)
1. Given $C(n, 2) = 66$, find n.
2. A bag contains 5 white balls, 8 black balls, 7 yellow balls, and 15 red balls. Find the probability that on two consecutive draws from this bag both balls will be red if the ball is not replaced after the first draw.
3. How many three-digit numbers can be formed with the digits 1, 2, 4, 6 if no digit is repeated?
4. Given $A = \{1, 2, 3, 4, 6\}$ and $B = \{2, 4, 6, 8\}$. Find the number of elements in $A \cup B$.
5. Given the set $S = \{1, 2, 3, 4, 5, 6\}$. Find the cross-partitions of $A_1 = \{1, 2, 3, 4\}$, $A_2 = \{5, 6\}$ and $B_1 = \{1, 2, 3\}$, $B_2 = \{4, 5, 6\}$.

Test #10. (Covers material in Chapter 10.)
1. Construct a truth table for $(p \to q) \wedge (q \to p)$.
2. State the converse, inverse, and contrapositive of the following implication: "If the sides of a triangle are all equal, then the triangle is equiangular."
3. Use a Veitch diagram to demonstrate that $(AB')' = A' + B = AB + A'$ for two general sets A and B.
4. Prove that for a two-valued algebra as defined in the text we have the identity $x(x + y) = x$.
5. Write the contrapositive of the converse of the implication, "if p, then q."

Part II. Special Examinations

A. An Examination on Graphing.
1. Plot the graphs of the following equations on rectangular coordinate paper:
   a) $y = 6 - x - x^2$
   b) $y = x^2 - 4x + 4$
   c) $y = \pm \sqrt{16 - 4x^2}$
   d) $y = x^3 + x - 4$
   e) $y = x^3 - x^2 + 7$
   f) $y = \pm \frac{1}{2}\sqrt{9 + x^2}$

g) $y = \pm \sqrt{4 + x^2}$ h) $y = \pm \frac{2}{3}\sqrt{x^2 - 9}$

i) $x^{\frac{1}{2}} + y^{\frac{1}{2}} = 2$

j) $\frac{1}{2}(3x - 10) - 3(1 - \frac{1}{4}y) = \frac{1}{4}(9y - 20)$

2. Plot the graphs of both loci on the same rectangular coordinate paper:

a) $\begin{cases} y = \frac{1}{2}(4x - 7) \\ y = 5 - 3x \end{cases}$ b) $\begin{cases} 2x^2 - y^2 = 4 \\ 3x^2 + 4y^2 = 12 \end{cases}$

c) $\begin{cases} 2x + 3y = 7 \\ xy + y^2 = 5 \end{cases}$ d) $\begin{cases} x^2 + 4y^2 = 16 \\ 16y^2 - x^2 = 16 \end{cases}$

e) $\begin{cases} x^2 + y^2 = 9 \\ 9x^2 + 16y^2 = 144 \end{cases}$ f) $\begin{cases} xy = 7 \\ 4x^2 - 9y^2 + 36 = 0 \end{cases}$

g) $\begin{cases} x^2 + 2xy = 5 \\ x^2 - xy + y^2 = 3 \end{cases}$ h) $\begin{cases} 4x^2 + 9y^2 = 36 \\ 3x + 2y = 0 \end{cases}$

i) $\begin{cases} xy = 6 \\ x^2 + y^2 = 25 \end{cases}$ j) $\begin{cases} xy = -12 \\ x^2 - 2y^2 + 2 = 0 \end{cases}$

k) $\begin{cases} x + y = 3 \\ xy = -3 \end{cases}$ l) $\begin{cases} x - \frac{3}{2}(y + 1) = 1 \\ y - \frac{1}{2}(x + 3) = 1 \end{cases}$

m) $\begin{cases} 4x^2 + y^2 = 61 \\ 2x - 1 = y \end{cases}$ n) $\begin{matrix} y^2 = 12x \\ 9x^2 - 16y^2 = 144 \end{matrix}$

o) $\begin{cases} x^2 - 10y = 19 \\ x^2 - 4y^2 = 5 \end{cases}$

3. Plot the graphs of the following equations:

a) $y(x + 1) = 2x$ b) $y = \sqrt{x^2 - 1}$

c) $y = x^4 - 2x^2$ d) $y^2 = x^3 - x^2$

e) $y(x^2 + x - 2) = x$ f) $x^2 - xy - 1 = 0$

g) $y = 2e^x$ h) $y = 2 \ln x$

i) $y = \frac{1}{2}(e^x + e^{-x})$ j) $xy = x - 2$

k) $y = x^5 - 9x^4 + 25x^3 - 15x^2 - 26x + 24$

B. A 1942 Honors Examination. (Time: 2 hours)

1. a) If r_1 and r_2 are the roots of $ax^2 + bx + c = 0$, what is the value of $r_1{}^2 r_2 + r_1 r_2{}^2$?

   b) Determine k so that there exists only one distinct pair of values satisfying the simultaneous equations:

$$y = 2x + k$$
$$y = x^2 - 5x + 6$$

2. a) Show that by adding 1 to the product of any 4 consecutive integers a perfect square is obtained.

   b) In the expansion of $(1 + x)^{43}$, the numerical coefficients of the $(2r + 1)$th and the $(r + 2)$th term are equal. Find r.

3. a) For what values of x is $\sqrt{x^2 - 2x - 15}$ real?

b) Prove that for any two distinct positive numbers a and b, their geometric mean ($\sqrt{ab}$) is less than their arithmetic mean.

4. a) Prove

$$\log\left(\frac{\sqrt{1 + x^2} - 1}{\sqrt{1 + x^2} + 1}\right) = 2[\log(\sqrt{1 + x^2} - 1) - \log x]$$

b) Given $\log 2 = 0.30103$. Find the number of zeros between the decimal point and the first significant figure when $(1/2)^{1000}$ is expressed as a decimal.

5. In a set of four numbers the first three are in geometric progression, the last three are in arithmetic progression with a common difference of 6, the first number is the same as the fourth. Find the numbers.

6. Three men, A, B, and C, set out at the same time to walk a certain distance. A walks 4 1/2 miles an hour and finishes the journey two hours before B, who walks one mile an hour faster than C and finishes the journey in three hours less time. What is the distance?

C. An Examination on Proofs.

Prove the following theorems or statements:
1. The Remainder Theorem.
2. The set of real numbers is dense.
3. If A and B are not disjoint, then $n(A \cup B) = n(A) + n(B) - n(A \cap B)$.
4. The square of an odd number is an odd number.
5. $\cosh x + \sinh x = e^x$
6. An equation of a straight line is of the first degree in x and y.
7. $\sqrt{2}$ is not a rational number.
8. If N is a real number, then $N^2 > 0$.
9. $|ab| = |a| \cdot |b|$
10. $1 + 3 + 5 + 7 + \ldots + (2n - 1) = n^2$

D. A True or False Examination.

Determine whether the following statements are true or false:
1. If $A = \{x | x^2 + x = 6\}$, then $2 \in A$.
2. The prime factors of 154 are 2, 7, and 11.
3. $a^2 x^3 = (ax)^5$
4. $|x| - |x + 2| + |- 3| = 1$ if $x \in A = \{x | x - 2 = 0\}$
5. $\{1, 3, 5, 7\} - \{2, 3, 4, 5\} = (1, 2, 4, 7)$
6. $A \cup \emptyset \cap \emptyset = A$
7. If $a < b$, then $a + c > b + c$.
8. The contrapositive of "if not-p, then not-q" is "if q, then p."
9. $(U \cup \emptyset)' = \emptyset$
10. $A \cap B = B \cap A$
11. If $a > b$ and $c < 0$, then $a/c < b/c$.
12. If $a \in A$, then $\{a\} \subset A$.

13. The set of integers is dense.

14. If a and b are positive real numbers and $a < b$, there is a positive integer n such that $na > b$.

15. There are infinitely many real numbers between any pair of distinct real numbers.

16. The multiplicative inverse of 7 is -7.

17. The set $\{(1, 2), (5, 2), (4, 2)\}$ describes a function.

18. If $f(x) = ax^2 + a$, then $f(2) = 5a$.

19. If $f(x) = x + 1$ and $g(x) = x - 1$, then $g[f(x)] = x$.

20. If $f(x) = 2x - 3$, then $[f(x + h) - f(x)] \div h = 2$.

21. The set of real numbers may form the domain of the function defined by $\{(x, y)|y = x^{-1}\}$.

22. The inverse of the function defined by $y = a - x$ is defined by $y = a - x$.

23. The slope of the line defined by $2y = 3x - 5$ is $3/2$.

24. The solution set of $x + y = 3$ and $x - y = 5$ is $(4, -1)$.

25. $\begin{vmatrix} 1 & -2 & 3 \\ 3 & -1 & -1 \\ 2 & -3 & 2 \end{vmatrix} = 10$

26. $\begin{pmatrix} 3 & 5 \\ 7 & 2 \end{pmatrix} + \begin{pmatrix} 2 & -2 \\ -2 & 2 \end{pmatrix} = \begin{pmatrix} 5 & 3 \\ 5 & 4 \end{pmatrix}$

27. If $7x + 3 < 2x + 8$, then $x < 1$.

28. The equation of a line through $P_1(4, 1)$ and $P_2(2, 5)$ is $y + 2x = 9$.

29. The inverse of $yx = k$ is $yx = k$.

30. $x + 1$ is a factor of $2x^4 - x^3 - 14x^2 - 5x + 6$.

31. The maximum value of $y = 16x - 4x^2$ occurs at $x = 2$.

32. The graph of $x^2 + 2y^2 = 1$ is a hyperbola.

33. $\{x|x^2 - x - 6 < 0\} = \{x|-2 \leq x \leq 3\}$

34. $\dfrac{6x - 4}{x^2 - 1} = \dfrac{1}{x - 1} + \dfrac{5}{x + 1}$

35. $\log \dfrac{x^2}{y} = 2 \log x + \log y$

36. $\log_2 4 = \log 4 - \log 2$

37. $\sinh(-x) = \sinh x$

38. $\lim\limits_{n \to \infty} \left(\dfrac{1}{n}\right) = 0$

39. The p-series is convergent if $p > 1$.

40. If $A \subset S$ and $B \subset S$ and $A \cup B = S$, then A and B are said to be exhaustive subsets.

41. If A and B are disjoint sets, then $n(A \cup B) = n(A) + n(B)$.

42. If $P(n, 2) = 6$, then $n = 3$.

43. The converse of a true implication is not always true.

44. If p and q are true, then $(p \wedge q) \rightarrow (p \vee q)$ is true.
45. If the domain of the function f defined by $y = [x]$ is $[1, 2[$, then the function is a constant.
46. The image of f defined by $3x - 6y = 2$ is a straight line.
47. If $x^4 + x^3 - 2x^2 + x + 3$ is divided by $x + 2$ the remainder is 1.
48. The distance between $P_1(-7)$ and $P_2(-4)$ is -3.
49. $a^2 > b^2$ if $a > b$.
50. Every repeating decimal expansion is a rational number.

E. An Examination on Computation.

A set of tables is necessary.

1. Compute $N = \dfrac{\sqrt[3]{0.9573} \, (3.21)^2}{98.32}$.

2. Find the solutions of $x^2 + x + 1 = 0$.

3. Find $\log_9 6$.

4. Compute $N = \sqrt{3\sqrt{72}}$.

5. Compute $N = x^x$ for $x = 1.1$.

6. Find the area inside the circle defined by the relation $x^2 + y^2 = 2$ and outside the right triangle which is inscribed in the circle and the hypotenuse of which equals the diameter of the circle and one leg of which is $\sqrt{2}$.

7. Find the value of $\sinh x + \cosh x$ for $x = 1.1$.

8. Find all the roots of $6x^4 + x^3 - 26x^2 - 4x + 8 = 0$.

9. Use determinants to find the solution set for the system
$$\begin{cases} 2x - 3y + 4z = 7 \\ x + y - z = 3 \\ 3x + 2y - 4z = 6 \end{cases}$$

10. Find the solution set for the system
$$\begin{cases} 3x^2 + xy + 2y^2 = 6 \\ 3x^2 + xy + 4y^2 = 9 \end{cases}$$

F. An Examination on Polynomials of the Second Degree.

1. If r_1 and r_2 are members of the solution set for the equation $ax^2 + bx + c = 0$, what is the value of $r_1{}^2 + 2r_1 r_2 + r_2{}^2$?

2. For what values of x is $(x^2 + 7x + 12)^{\frac{1}{2}}$ real?

3. Sketch the graph of $9x^2 - 16y^2 - 18x + 96y + 9 = 0$.

4. Find the characteristic points of the parabola $y^2 - 4y - 4x + 8 = 0$.

5. Find the inverse of the function f defined by $y = 1/2 \sqrt{16 - x^2}$ over the domain $-4 \leq x \leq 0$.

6. The perimeter of a right triangle whose hypotenuse is 25 is 56. Find the other sides of the triangle.
7. Find the values of x for which $1/x > 1 - x$.
8. Given the relation $y^2 + x^2 = 1$. Let x change by Δx and y correspondingly by Δy. Find the fraction $\Delta y / \Delta x$ and let $\Delta x = 0$ in the result. (If $\Delta x \to 0$, then $\Delta y \to 0$.)
9. For what values of k will the roots of $x^2 - kx + 4 = 0$ differ by 3?
10. Draw the graph of $6x^2 - 7xy - 3y^2 - 4x + 6y = 0$.

G. An Examination on Definitions.

Define and give an example of each of the following mathematical terminologies:

1. The converse of an implication.
2. A proper subset.
3. The complement of a set.
4. Two disjoint sets.
5. The intersection of two sets.
6. A rational number.
7. The absolute value of a real number.
8. The conjugate of a complex number.
9. A function relating the sets X and Y.
10. The inverse function.
11. A second-order determinant.
12. The equality of two matrices.
13. The inverse of a square matrix.
14. A quadratic equation in one variable.
15. The vertical asymptote of the graph of a function defined by $y = f(x)$.
16. An exponential function.
17. A monotone increasing function.
18. An arithmetic progression.
19. The limit of a sequence.
20. $n!$
21. A partition of a set.
22. A permutation.
23. Mutually exclusive events.
24. A tautology.
25. A group.

H. An Examination on Symbols.

Give a word description and an example of each of the following symbols.

1. $\to$ 2. $\vee$ 3. Δx 4. $|n|$ 5. (x, y)
6. $P(n, r)$ 7. $\pm$ 8. $>$ 9. $\sqrt{n}$ 10. $\leftrightarrow$
11. $A \cup B$ 12. $\emptyset$ 13. $\{\ldots\}$ 14. A' 15. $a \in A$
16. $1.\overline{3}$ 17. $-a$ 18. $f(x)$ 19. $\sim p$ 20. $f: X \to Y$

21. $[x]$ 22. f^{-1} 23. $D_x y$ 24. $n!$ 25. $\ln N$

26. $\sum\limits_{i=1}^{n}$ 27. $\begin{pmatrix} x & y \\ r & s \end{pmatrix}$ 28. $\begin{vmatrix} x & y \\ r & s \end{vmatrix}$ 29. $\sinh x$ 30. $\lim\limits_{n \to \infty} a_n$

I. An Examination on Symbolic Problems.

1. $\{1, 3, 5\} \cup \{2, 3, 4\} =$ 2. $B \cap B' =$
3. $[7, 11] \cup [8, 12[=$ 4. $A \cap (B \cup B') =$
5. $A = \{x | x + 3 = 7\} =$ 6. $\sim (\sim p) \leftrightarrow$
7. $(5 + 2i) \div (2 - 3i) =$ 8. $U - A' =$
9. $f(x) = x^2 - x - 56 \to f(1) =$ 10. $(1 - i)^7 =$
11. If $B \subset A$, then $A \cap B' =$ 12. $1.\overline{45} =$

13. $\begin{vmatrix} 1 & -5 & 3 \\ 2 & 1 & 0 \\ 3 & -1 & -2 \end{vmatrix} =$ 14. $\begin{pmatrix} 3 \\ 5 \\ 2 \end{pmatrix} - \begin{pmatrix} -3 \\ -5 \\ -2 \end{pmatrix} =$

15. $A = \begin{pmatrix} 3 & -1 \\ 2 & 1 \end{pmatrix} \to A^{-1} =$ 16. $A \cup \emptyset =$

17. $3x + 7 > 0 \to x >$ 18. $x^3 = 1 \to x =$
19. $x^3 + 2x^2 - 23x - 60 = 0 \to x =$ 20. $P(7, 3) =$

J. An Examination on Originality. (Open Book)

This examination deals with the development of some properties of quaternions, a hypercomplex number system, which was not discussed in the text. We shall state some definitions and notation and then ask you to prove some properties based on the given information.

Let

$$ i = \begin{pmatrix} \sqrt{-1} & 0 \\ 0 & -\sqrt{-1} \end{pmatrix} \qquad j = \begin{pmatrix} 0 & 1 \\ -1 & 0 \end{pmatrix} $$

$$ k = \begin{pmatrix} 0 & \sqrt{-1} \\ \sqrt{-1} & 0 \end{pmatrix} \qquad I = \begin{pmatrix} 1 & 0 \\ 0 & 1 \end{pmatrix} $$

1. Show that
   a) $i^2 = j^2 = k^2 = -I$
   b) $ij = k, jk = i,$ and $ki = j$
   c) $ji = -k, kj = -i,$ and $ik = -j$
   d) $iij = -j, ijj = -i,$ and $kki = i$
   e) $k(-kj) = j$ and $j(-ji) = i$

   Definition. *Any linear combination of* I, i, j, k *is called a* **quaternion**,

   $$ q = aI + bi + cj + dk $$

   where a, b, c, d $\in$ R.

2. Show that $q = \begin{pmatrix} x & y \\ -\bar{y} & \bar{x} \end{pmatrix}$

where x and y are the complex numbers

$$x = a + b\sqrt{-1} \qquad \bar{x} = a - b\sqrt{-1}$$

$$y = c + d\sqrt{-1} \qquad \bar{y} = c - d\sqrt{-1}$$

3. Show that the set of quaternions is closed under multiplication; i.e.,

$$q_1 q_2 = \begin{pmatrix} V_1 & V_2 \\ V_3 & V_4 \end{pmatrix}$$

where $V_3 = -\bar{V}_2$ and $V_4 = \bar{V}_1$.

Definition. *The **conjugate** of the quaternion $q = aI + bi + cj + dk$ is defined to be $\bar{q} = aI - bi - cj - ck$.*

4. Show that

$$\bar{q} = \begin{pmatrix} \bar{x} & -y \\ \bar{y} & x \end{pmatrix}$$

5. Show that $q\bar{q} = nI$ where $n = x\bar{x} + y\bar{y}$.

Definition. *The expression $n(q)$ is called the **norm** of q where $n = x\bar{x} + y\bar{y}$.*

6. Show that the conjugate of the product of two quaternions is equal to the product of their conjugates taken in reverse order; i.e.,

$$\overline{(q_1 q_2)} = \bar{q}_2 \bar{q}_1$$

7. Show that the norm of a product is the product of the norms; i.e.,

$$(q_1 q_2)(\bar{q}_2 \bar{q}_1) = n(q_1) n(q_2) I$$

8. Show that the commutative axiom for addition holds for quaternions.
9. Show that the commutative axiom for multiplication does not hold for quaternions.

Definition. *The quaternion*

$$q_0 = \begin{pmatrix} 0 & 0 \\ 0 & 0 \end{pmatrix}$$

*is called the **identity element** for addition.*

Definition. *The quaternion*

$$q = \begin{pmatrix} -x & -y \\ \bar{y} & -\bar{x} \end{pmatrix} = -q$$

*is called the **additive inverse** of q.*

10. Show that quaternions form an additive commutative group.

Part III. Final Examinations
(Time: 3 hours each)

Exam. #1.

Part A. True or False?

1. $X \cup Y$ is the set of elements which belong to either X or Y or both X and Y.
2. Set A is a proper subset of B if and only if A is a subset of B.
3. The null set is the set which contains only one element.
4. The real number $(-a)$ is called the multiplicative inverse of the real number a.
5. If a is negative, then $|a| = a$.
6. The commutative law of multiplication states that $a(b + c) = ab + ac$.
7. Let x and y be positive real numbers. Then $x(-y) = -xy$.
8. The number $\sqrt{5}$ is irrational.
9. The length of the segment between $P_1(x_1, a)$ and $P_2(x_2, a)$ is given by $|x_2 - x_1|$.
10. The complex numbers $a + bi$ and $c + di$ are equal if $a = c$ and $b = d$.
11. $x^2 y^3$ times axy^3 equals $ax^3 y^6$.
12. If $a > b$ and $c < 0$, then $bc > ac$.
13. If $a < b$, then $b + c > a + c$.
14. The domain of a relation R is the subset $X: \{x | x$ is the first element of at least one of the pairs (x, y) of $R\}$
15. The relation $\{(x, y) | y = x^2\}$ is also a function.
16. If the domain of $f(x) = x$ is limited to $[-1, 3]$, then the range is $[0, 3]$.
17. The rational function

$$y = \frac{P(x)}{Q(x)}$$

is defined everywhere.

18. If $a > b$ and $b > c$, then $c - x < a - x$.
19. If $a = 0$ and $b \neq 0$, then the solution of $ax^2 + bx - c = 0$ is $x = cb^{-1}$.
20. $\{x | x^2 - 5x + 6 = 0\} = \{2, 3\}$
21. $(\sqrt{-16})(\sqrt{-25}) = 20$
22. $a^{-1} + b^{-1} = (a + b)^{-1}$
23. For any real number a, $\sqrt{a^2} = |a|$
24. $\dfrac{k + a}{k} = a$
25. The binomial coefficients $\dbinom{n}{r} = \dbinom{n}{n-r}$

Part B. Answer seven of the following questions.

1. Find all the roots of $x^5 - 2x^4 - 2x + 4 = 0$.

2. Given $f: x^2 + y = 0$ and $d_f: 0 \leq x \leq 3$.
   a) Sketch the graph of f.
   b) Find f^{-1} and its domain.
   c) Sketch the graph of f^{-1}.

3. Eight coins are tossed. What is the probability that exactly three of them are heads?

4. State and prove the Factor Theorem for polynomials.

5. Find the domain for which $\log (15 - 5x - x^2)$ is positive.

6. Use a Veitch diagram to demonstrate that $A + BC = (A + B)(A + C)$ for three general sets A, B, and C.

7. Find the extrema of the linear function $f: 2x - 2y + 3$ defined over the convex polygon which is the solution set of the system

$$4x - y + 1 > 0$$
$$2x - 3y - 7 < 0$$
$$x + y - 6 < 0$$

8. Find the sets resulting from each of the following operations:
   a) $[1, 7] \cup [-1, 3]$
   b) $\{1, 6, 2\} \cap \{2, 4, 6, 8\}$
   c) If $U = \{$an integer $x | 0 \leq x \leq 10\}$, find A' if $A \subset U = \{x | x < 6\}$.
   d) $A \times B$ if $A = \{2, 4\}$ and $B = \{0, 1\}$
   e) $A \cap (B \cup B')$

9. Find the solution set for the system

$$\begin{cases} 3x + 2y - z - 2w = 9 \\ x - 2y + 6z + w = 6 \\ 2x + y - z + 3w = -3 \\ -3x + 2y + 3z + w = -5 \end{cases}$$

10. Find the first six terms of the binomial series for $(1 - 2x)^{\frac{1}{2}}$.

Exam. #2.

Part A. True or False?

1. If a and b are integers, then a/b is an irrational number.
2. If $a < b$, then $b + c < a + c$ for all c.
3. $|a| = a$ if $a < 0$.
4. If $a > b$ and $c < 0$, then $bc > ac$.
5. If A^{-1} is the inverse of the matrix A, then AA^{-1} is the determinant of A.
6. The multiplication of square matrices is commutative.
7. The ordered pairs $(1, 2)$ and $(1, -2)$ define a function.
8. If $y = f(x) = x - 5$, then $f^{-1}(x) = x + 5$.
9. If $f(x) = x$ and $g(x) = x^2$, then $d_{f/g}$ is R.
10. If $ab > 0$, then either $a > 0$ and $b > 0$ or $a < 0$ and $b < 0$.
11. If $x = 3$, then $2x^2 - 7x + 1 < 0$.

12. $\begin{vmatrix} 12 & -16 \\ 15 & -20 \end{vmatrix} = 0$

13. The relation $y = f(x) = x$ is a function such that $f^{-1} = f$.

14. If $2^x = 3$, then $\log_2 3 = x$.

15. The null set is the set which contains only one element.

16. $(3 - i)^{-1} = .3 + .1i$

17. $(x^2 y^3)(- 2xy^{-1}) = - 2x^3 y^2$

18. $A' \cap B' = (A \cap B)'$

19. If $a = - 5$ and $b = 3$, then $|a| > |b|$.

20. If $a > b > c > d$, then $a - x > d - x$.

21. $y = e^x$ is a monotone increasing function.

22. In an A.P., $S_n = (n/2)(a + l)$.

23. $3 - 2i + 4 = 4i + 7 - 6i$

24. If $2x^2 - 3x - 5 = 0$, then the sum of the roots is 1.5.

25. The distance between $P_1(x_1, a)$ and $P_2(x_2, a) = |x_2 - x_1|$.

Part B.

1. If $C(n, n - 2) = 66$, find n.
2. Find all the roots of $3x^4 - 11x^3 - 68x^2 - 17x + 21 = 0$.
3. Write the series expansion of $\sqrt{1 + 2x}$.
4. Decompose into partial fractions:

$$\frac{2y^2 - 2y + 1}{y^2(y - 1)^2}$$

5. Prove that every rational integral equation of the nth degree has exactly n roots.

6. Solve the system

$$\begin{cases} y - 4 < 0 \\ 3x + 2y + 6 > 0 \\ 7x - 4y - 12 < 0 \end{cases}$$

7. Sketch the graph of $9x^2 - 16y^2 - 36x - 32y - 124 = 0$.
8. Find the rational number represented by $12.\overline{5}$.
9. Prove that $1 + 3 + 5 + 7 + \ldots + (2n - 1)^2 = n^2$.
10. Prove that $1 > 0$.

Exam. #3.

1. Prove that the additive inverse of the additive inverse of an element is the element itself.
2. Find $[(1 + i)^2]^3$.
3. Solve the system

$$\begin{cases} 3x + y - z = 15 \\ x + 3y - z = 17 \\ x + y - 3z = - 7 \end{cases}$$

4. What is the probability of throwing a 4 with three dice?

5. State and prove the Remainder Theorem for polynomials.

6. Express as a single logarithm

$$\frac{1}{2} \log 8 - \log 6 + \frac{1}{3} \log 54 - \frac{1}{6} \log 2$$

7. Find the first four terms of the series expansion for $(1 - x)^{\frac{1}{2}}$ and use these to approximate the value of $\sqrt{1/2}$ to two decimal places.

8. Find the set resulting from each of the following operations:
   a) $\{(x, y)|x^2 - y^2 = 8\} \cap \{(x, y)|x - y = 2\}$
   b) $U = \{1, 2, 3, 4, 5, 6, 7\} \cap A'$ if $A = \{2, 4, 6\}$

9. Prove the following: "The rate of change of a linear function is equal to the slope of its straight line."

10. Find the inverse of $f: y = \sqrt{16 - x^2}, d_f: -4 \le x \le 0$.

ANSWERS TO THE EXERCISES

Exercise I.

1. a) 6   **b)** 1   **c)** x **d)** 1   **e)** $a^4 - a^3$

2. a) 0   **b)** 3   **c)** -3 **d)** 0   **e)** 2

3. a) $-6a^3x^3y^2$ **b)** $-8a^3b^6$ **c)** $3x^{10}y^2$ **d)** $\dfrac{x^{4n-1}}{y^{m+3}}$

     e) $\sqrt{x-y}$ **f)** $7\sqrt{2}$

4. a) $x^4 - 4x^3y + 4x^2y^2 - 7xy^3 - 6y^4$

     b) $2x^3 - 2x^3y + x^2y + x^2y^2 + xy - 2xy^2 + y^2$

     c) $x^4 - 4x^3y + 6x^2y^2 - 4xy^3 + y^4$

     d) $7 + i$ **e)** $4.48x^2 - 12.20xy + 4.48y^2$

5. a) $(x + y)(x + z)$ **b)** $2x(x + y)$ **c)** $(y + 4)(y - 3)$

     d) $(a - 3b)(a^2 + 3ab + 9b^2)$ **e)** $(x + 13)(x - 7)$

     f) $(x - 5)(x - 2)$ **g)** $(3x - y)(x - 3y)$

     h) $(3x + 2y)(2x + 5y)$ **i)** $(2ab - 3x)^3$

     j) $(3a - b)(3x - 2y)(9x^2 + 6xy + 4y^2)$

6. a) $x^{\frac{1}{2}} + 1$ **b)** $x^2 + 3xy + 9y^2$ **c)** $4x^2 - 3xy + 2y^2$

7. a) $\dfrac{2y}{x^2 - y^2}$ **b)** $\dfrac{4b}{a^2 - b^2}$ **c)** $\dfrac{x^2 + 3x - 5}{(x - 3)(x + 2)(x - 1)}$

     d) $\dfrac{28y^2 - 12x^2}{(3x + 2y)(2x + 5y)(x - y)}$ **e)** $\left(\dfrac{x + a}{x - a}\right)^2$

     f) $\dfrac{ay + bx}{ay - bx}$ **g)** $\dfrac{15}{16}$ **h)** 0

8. a) $a - b$ **b)** $\sqrt{3x} + 3\sqrt{3y} - 3\sqrt{xy} - 9y$

     c) $2\sqrt{x} - 2\sqrt{2x} + 2x\sqrt{2} - 4x$

9. a) $\dfrac{3\sqrt{2}}{2}$ **b)** $\dfrac{2^{\frac{5}{6}}}{4}$ **c)** $-\dfrac{1}{3}(\sqrt{6} + \sqrt{2} + \sqrt{15} + \sqrt{5})$

     d) $\dfrac{(\sqrt{x} + \sqrt{y})^2}{(x - y)}$ **e)** $\dfrac{1}{5a + 13b}(13b - 7a - 13\sqrt{a^2 - b^2})$

     f) $\dfrac{6 - 5\sqrt{x} - 6x}{4 - 9x}$

239

10. a) $2\sqrt{6}$ **b)** $\dfrac{1}{3} - \dfrac{2\sqrt{2}}{3}i$ **c)** $5 + 5i$

d) $\dfrac{1}{3}[2\sqrt{2} - 1 + (2 + \sqrt{2})i]$

Exercise II.

1. a) $\{1, 2, 3, 4, 5, 6, 7, 8, 9, 10, 11\}$
  b) $\{2, 4, 6, 8, 10, \ldots\} = \{2n,$ where n is a positive integer$\}$
  c) $\{$Jack, Joe, Jim, Jake$\}$
  d) $\{2, 3, 5, 7, 11, 13, 17, 19\}$
  e) $\{6, 7, 8, 9, 10, 11, 12, 13, 14, 15, 16\}$
  f) $\{a, e, i, o, u\}$
  g) $\{7\}$
  h) $\emptyset$

2. a) $\{1, 3\}, \{1, 5\}, \{3, 5\}$ **d)** $\{3\}$
  b) $\{2, 4\}$ **e)** $\emptyset$
  c) $\{a\}$ **f)** $\{$Joe, Jake$\}$

3. a)

a	b	c	d	e	f	$\ldots$	z
1	2	3	4	5	6	$\ldots$	26

b)

1	2	3	4	$\ldots$
-1	-2	-3	-4	$\ldots$

c)

1	2	3	4	5	6	7	$\ldots$
2	4	6	8	10	12	14	$\ldots$

4. a) $\{1, 2, 3, 4\}$ **b)** $\{1, 2, 5, 6\}$
  c) $\{1, 2, 3, 4\}$ **d)** $\{2, 3\}$
  e) $\{2, 4\}$ **f)** $\emptyset$
  g) $\{2, 3, 4\}$ **h)** $\emptyset$
  i) U **j)** $\emptyset$
  k) $\{x \mid x$ is an integer$\}$

5. a) A **b)** B **c)** $U - (A - B)$ **d)** $\emptyset$ **e)** U **f)** $A - B$ **g)** B' **h)** A'

6. Figure 66

7. a) $\{(1, a), (1, b), (2, a), (2, b), (3, a), (3, b)\}$
  b) $\{(x, c), (x, d), (y, c), (y, d), (z, c), (z, d)\}$
  c) $\{(x_1, y_1), (x_1, y_2), (x_2, y_1), (x_2, y_2)\}$
  d) (Chevrolet, coupe), (Chevrolet, sedan), (Chevrolet, convertible), (Ford, coupe), (Ford, sedan), (Ford, convertible)
  e) (true, true), (true, false), (false, true), (false, false)

9. a) $\{1, 3, 5, 7, 9\}$ **b)** $\{1, 2, 8, 9\}$
  c) $\{1, 2, 3, 4, 5, 6, 7, 9\}$ **d)** $\{2, 4, 6, 8\}$
  e) $\{5, 6, 7\}$

11. T, T, T, T, T, T, T, F, F, F, F, T, F, T, T

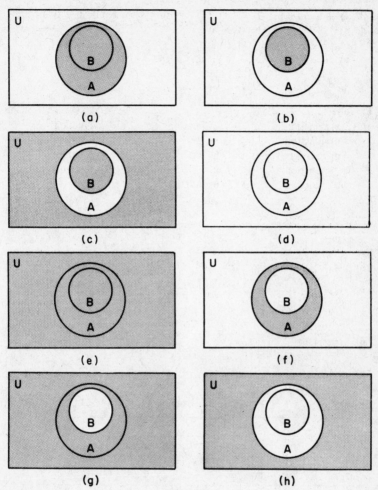

Fig. 66

Exercise III.

1. $-1, -3, 6, -\dfrac{1}{3}, \sqrt{2}$

2. $1, \dfrac{1}{7}, 3, -\dfrac{1}{3},$ does not exist, $\dfrac{1}{\sqrt{2}}$

4. T, F, F, T, T, T, F, T, T, T

Exercise IV.

1. **a)** $\dfrac{4}{3}$ **b)** $\dfrac{8}{3}$ **c)** $\dfrac{27}{110}$ **d)** $\dfrac{28}{9}$ **e)** $\dfrac{1457}{1100}$ **f)** $\dfrac{113}{9}$

3. a) 4   **b)** 10   **c)** 8   **d)** 7   **e)** 3   **f)** 12

4. a) $[18, 23]$ **b)** $]1, 5]$ and $[6, 9[$ **c)** $[7, 18[$

     d) $[2, 4]$ **e)** $[1, 6]$ **f)** $] -\infty, 2[\cup]4, +\infty[$

5. a) $10 + 3i$ **b)** $-5 + 5i$ **c)** $9 + 7i$

     d) $4 + 19i$ **e)** $\dfrac{23}{13} + \dfrac{11}{13}i$ **f)** $-\dfrac{13}{5} - \dfrac{14}{5}i$

Exercise V.

H **1. a)** function   **b)** relation   **c)** function

     d) function   **e)** function   **f)** function

     g) relation   **h)** function   **i)** relation

     j) function   **k)** function   **l)** function

     m) function   **n)** function

2.

	x	-3	-2	-1	0	1	2	3
a)	$f(x)$	10	5	2	1	2	5	10
b)	$f(x)$	0	1	2	3	4	5	6
c)	$f(x)$	6	5	4	3	2	1	0
d)	$f(x)$	$1 + \sqrt{3}i$	$1 + \sqrt{2}i$	$1 + i$	1	2	$1 + \sqrt{2}$	$1 + \sqrt{3}$
e)	$f(x)$	$-\dfrac{1}{4}$	$-\dfrac{1}{3}$	$-\dfrac{1}{2}$	-1	—	1	$\dfrac{1}{2}$
f)	$f(x)$	9π	4π	π	0	π	4π	9π

4.

	a)	**b)**	**c)**	**d)**	**e)**	**f)**
$f + g$	$x^2 + x - 3$	$\dfrac{x^2}{x + 1}$	$\dfrac{(x+1)^3 + 1}{(x+1)^2}$	$2x$	$1^x + x + 1$	$a + b$
$f - g$	$x^2 - x - 3$	$\dfrac{2 - x^2}{x + 1}$	$\dfrac{1 - (x+1)^3}{(x+1)^2}$	2	$1^x - x - 1$	$a - b$
fg	$x^3 - 3x$	$\dfrac{x - 1}{x + 1}$	$\dfrac{1}{x + 1}$	$x^2 - 1$	$1^x(x + 1)$	ab
$\dfrac{f}{g}$	$\dfrac{x^2 - 3}{x}$	$\dfrac{1}{x^2 - 1}$	$\dfrac{1}{(x+1)^3}$	$\dfrac{x + 1}{x - 1}$	$\dfrac{1^x}{x + 1}$	$\dfrac{a}{b}$

5. **a)** $z = x^2 - x$ **b)** $z = x^4$ **c)** $z = |x|$

6. **a)** $y = \dfrac{1}{3}(x + 2)$ **b)** $y = x - 5$ **c)** $y = x$

   d) $y = x - k$ **e)** $y = \dfrac{x}{c}$ **f)** $y = 3 - x$

   g) $ay + bx = -c$ **h)** $my = x - b$

7. **a)** r_f: the set of integers   **b)** $r_f: \{y|y = n^2\}$

8. $f(a) = a + 1, f(x + h) = x + h + 1, f(x - h) = x - h + 1,$

   $f(x + h) - f(x) = h, \dfrac{f(x + h) - f(x)}{h} = 1$

9. **a)** $x \in R, x \neq 0$ **b)** $[0, \infty[$ **c)** $x \in R, x \neq 1, x \neq 2$

10. $d_{f+g} = R, d_{f-g} = R, d_{fg} = R, d_{\frac{f}{g}} = R\,(x \neq 0)$

Exercise VI.

1. **a)** $y \in R$ **b)** $\dfrac{3}{4} \leq y \leq \dfrac{15}{4}$ **c)** $6 \leq y \leq 8$

   d) $3 \leq y \leq 6$ **e)** $10 \leq V \leq 330$

2. **a)** $m = 2, a = -\dfrac{5}{4}, b = \dfrac{5}{2}$ **b)** $m = -\dfrac{1}{2}, a = \dfrac{1}{3}, b = \dfrac{1}{6}$

   c) $m = -1, a = b = 7$ **d)** $m = 2, a = -4, b = 8$

   e) $m = -6, a = 4, b = 24$ **f)** $m = \dfrac{1}{2}, a = \dfrac{7}{5}, b = -\dfrac{7}{10}$

3. **a)** -1 **b)** -1 **c)** 1 **d)** $-\dfrac{1}{3}$ **e)** 1 **f)** -3

   g) 0 **h)** undefined

4. **a)** $\left(\dfrac{7}{4}, \dfrac{1}{4}\right)$ **b)** $\left(\dfrac{9}{5}, \dfrac{1}{5}\right)$ **c)** $\left(-\dfrac{30}{a}, -\dfrac{60 + 3a}{ab}\right)$

   d) $\left(-\dfrac{1}{2}, \dfrac{7}{4}\right)$ **e)** $x = \dfrac{7}{3}, y = \dfrac{10}{3}, z = -\dfrac{14}{3}$

   f) $x = -3, y = \dfrac{19}{4}, z = \dfrac{35}{4}$ **g)** $x = 2, y = 3, z = -1$

   h) $x = -19, y = -7, z = 10$

5. **a)** $\begin{pmatrix} 4 & -5 \\ 5 & 1 \end{pmatrix}$ **b)** $\begin{pmatrix} -4 & 2 \\ 1 & 1 \\ -3 & -7 \end{pmatrix}$

   c) cannot be added

   d) $\begin{pmatrix} -1 & 1 & 2 \\ -1 & -7 & 1 \end{pmatrix}$ **e)** $\begin{pmatrix} 2 & -2 \\ 12 & 0 \end{pmatrix}$

   f) $\begin{pmatrix} 2a_1 + a_2 & 2b_1 + b_2 \\ 2c_1 + c_2 & 3d_1 \end{pmatrix}$ **g)** $\begin{pmatrix} 7 & -1 \\ 1 & 4 \end{pmatrix}$

   h) $\begin{pmatrix} 5 & 1 \\ 4 & 5 \\ 10 & 14 \end{pmatrix}$ **i)** (2)

j) $\begin{pmatrix} 24 & 0 & 15 \\ 9 & 3 & 6 \\ 12 & 39 & 9 \end{pmatrix}$ k) $\begin{pmatrix} 5 & 7 & 5 \\ -6 & 7 & -6 \end{pmatrix}$

l) $\begin{pmatrix} 51 & 0 & 17 & -34 \\ 33 & 0 & 11 & -22 \end{pmatrix}$

6. a) $X = \begin{pmatrix} -3 \\ -3 \end{pmatrix}$ **b)** $X = \begin{pmatrix} -\dfrac{1}{15} \\ -\dfrac{2}{15} \end{pmatrix}$ **c)** $X = \begin{pmatrix} \dfrac{8}{23} \\ -\dfrac{19}{23} \end{pmatrix}$

7. a) $x > -\dfrac{7}{3}$ **b)** $x > 5$ **c)** $x \geq 8$

d) $x \leq 6$

8. Figure 67

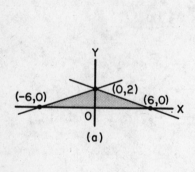

(a)

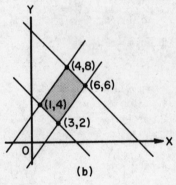

(b)

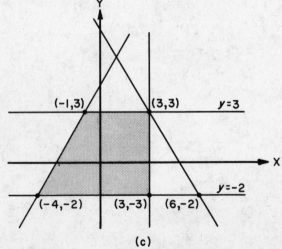

(c)

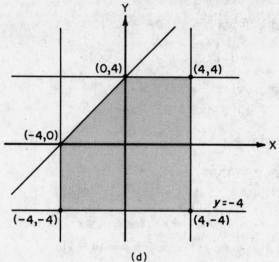

(d)

Fig. 67

9(a). **a)** $f_{max}(6, 0) = 51; f_{min}(-6, 0) = -45$
b) $f_{max}(3, 2) = 3; f_{min}(4, 8) = -61$
c) $f_{max}(3, -3) = 63; f_{min}(-1, 3) = -41$
d) $f_{max}(4, -4) = 83; f_{min}(0, 4) = -45$

9(b). **a)** $f_{max}(6, 0) = 0.45; f_{min}(-6, 0) = -13.95$
b) $f_{max}(4, 8) = 18.05; f_{min}(3, 2) = 1.85$
c) $f_{max}(3, 3) = 4.35; f_{min}(-4, -2) = -16.55$
d) $f_{max}(4, 4) = 8.05; f_{min}(-4, -4) = -21.55$

10. 3 and 7

Exercise VII.

1. **a)** 0, 2 **b)** $\pm\sqrt{17}$ **c)** 3, −2 **d)** $-\frac{1}{4} \pm \frac{\sqrt{61}}{4}$

e) $-\frac{1}{2} \pm \frac{\sqrt{3}}{2}i$ **f)** $-\frac{3}{4} \pm \frac{\sqrt{41}}{4}$ **g)** $-\frac{2}{3}$ (twice)

h) $\frac{1}{2} \pm \sqrt{3}$ **i)** $1 \pm \frac{\sqrt{6}}{3}i$ **j)** $-1 \pm \sqrt{2i}$

2. **a)** (3, 0), (6, −2) **b)** (5, 3), (2, 6)

3. **a)** (4, 1), (−4, −1), (2, 2), (−2, −2)

b) $\left(\frac{2}{3}, \frac{9}{4}\right), \left(-\frac{2}{3}, -\frac{9}{4}\right), \left(\frac{3}{2}, 1\right), \left(-\frac{3}{2}, -1\right)$

4. **a)** ($\pm i, \pm 2$) **b)** ($\pm 2i, \pm 3i$)

5. **a)** $\left(0, -\frac{5}{2}\right), (3, 2), (-3, 2)$ **b)** $\left(\pm\sqrt{54}, \frac{7}{2}\right), (\pm 3, -1)$

6. a) $\left(\pm\dfrac{1}{3}\sqrt{6}, \pm\dfrac{1}{2}\sqrt{6}\right), \left(\pm\dfrac{1}{2}\sqrt{6}, \mp\dfrac{1}{2}\sqrt{6}\right)$

   b) $\left(\pm\dfrac{1}{2}, \mp 1\right), \left(\pm\sqrt{2}, \pm\dfrac{\sqrt{2}}{3}\right)$

   c) $\left(\pm\dfrac{13}{51}\sqrt{51}, \pm\dfrac{8}{51}\sqrt{51}\right), (\pm 5, \mp 3)$

7. a) $(-2, -1), (-1, -2), (3, -1), (-1, 3)$

   b) $(2, 1), (1, 2), \left(\dfrac{-11 + \sqrt{209}}{6}, \dfrac{-11 - \sqrt{209}}{6}\right),$

   $\left(\dfrac{-11 - \sqrt{209}}{6}, \dfrac{-11 + \sqrt{209}}{6}\right)$

9. a) $f^{-1}: y = 2\sqrt{x}$ with domain $0 \le x \le 4$

   b) $f^{-1}: y = -\sqrt{16 - x^2}$ with domain $0 \le x \le 4$

   c) $f^{-1}: y = -\dfrac{3}{2}\sqrt{4 - x^2}$ with domain $[0, -2]$

   d) $f^{-1}: y = \sqrt{x^2 + 9}$ with domain $[0, 4]$

Exercise VIII.

2. a) $a_1 = -2, a_2 = -8$　　　　**b)** $a_1 = 0, a_2 = -6$
   c) $a_1 = -2, a_2 = -1$　　　　**d)** $a_1 = -13, a_2 = 17$
   e) $a_1 = \dfrac{17}{2}, a_2 = -\dfrac{77}{2}$

3. a) $1, 2, -3$　　**b)** $1, \dfrac{1}{2}, \dfrac{1}{2}$　　**c)** $3, -4, 5$

   d) $\dfrac{1}{2}, \dfrac{-1 \pm \sqrt{3}}{2}$　　**e)** $0, \dfrac{3}{2}, -\dfrac{1}{3}$

   f) $1, -3, -\dfrac{1}{2} \pm \dfrac{\sqrt{3}}{2}i$　　**g)** $-2, -3, \dfrac{1}{2} \pm \dfrac{\sqrt{7}}{2}i$

   h) $1, -2, 4$　　**i)** $1, 3, 1 \pm i$

   j) $1, 2, -2, 4$　　**k)** $-3, \dfrac{1}{2} \pm \dfrac{\sqrt{3}}{2}i, \dfrac{1}{2}$

   l) $-1, \pm i$　　**m)** $-1, 2, 5, -5$

   n) $-\dfrac{3}{2}, -\dfrac{3}{2}, \dfrac{1}{2} \pm \dfrac{\sqrt{5}}{2}$　　**o)** $-\dfrac{1}{2}, -\dfrac{3}{4}, \dfrac{3}{2}, -\dfrac{1}{2} \pm \dfrac{\sqrt{3}}{2}i$

   p) $1, -\dfrac{1}{2} \pm \dfrac{\sqrt{3}}{2}i$　　**q)** $1, 2, -\dfrac{1}{3}, \dfrac{3}{2}, \dfrac{-1 \pm \sqrt{7}i}{4}$

4. a) $\left(\dfrac{1}{3}, -\dfrac{131}{27}\right), (1, -5)$

   b) $\left(\dfrac{2\sqrt{3}}{3}, \dfrac{-16\sqrt{3} - 36}{9}\right)\left(-\dfrac{2\sqrt{3}}{3}, \dfrac{16\sqrt{3} - 36}{9}\right)$

c) $(1, 1)$, $\left(-\dfrac{1}{3}, \dfrac{91}{27}\right)$ d) $(1, 3)$

e) at $x = \dfrac{1}{2}, \dfrac{1}{6} \pm \dfrac{\sqrt{13}}{6}$

6. a) $x > 2, x < -3$ b) $x \geq 1, x \leq -3$ c) $1 \leq x \leq 3$
   d) $-4 < x < 4$ e) $x \geq -1$ f) $-7 < x < -1$

8. a) $\dfrac{1}{x + 1} + \dfrac{2}{2x + 3} - \dfrac{3}{2x - 3}$

b) $3x + \dfrac{2}{x + 1} + \dfrac{3}{x - 1} - \dfrac{4}{x + 2}$

c) $\dfrac{1}{x^2} + \dfrac{1}{(x - 1)^2}$

d) $\dfrac{2}{x - 1} + \dfrac{5}{x - 3} - \dfrac{5}{x - 2}$

e) $\dfrac{2}{x - 1} - \dfrac{1}{x^3} - \dfrac{1}{x^2} - \dfrac{2}{x}$

f) $1 + \dfrac{2}{x - 1} + \dfrac{2}{(x - 1)^2}$

g) $\dfrac{7}{8(x + 1)} - \dfrac{1}{8(x - 1)} - \dfrac{3x - 4}{4(x^2 + 1)} - \dfrac{3x - 4}{2(x^2 + 1)^2}$

h) $\dfrac{1}{x - 1} + \dfrac{2x - 3}{x^2 + x + 1}$

i) $\dfrac{1}{x + 2} - \dfrac{1}{x^2 + x + 1} + \dfrac{2x - 3}{(x^2 + x + 1)^2}$

j) $\dfrac{x + 2}{x^2 + 1} - \dfrac{x + 3}{(x^2 + 1)^2}$

k) $\dfrac{1}{x} - \dfrac{x}{x^2 + 1}$

l) $\dfrac{4}{x^2} - \dfrac{1}{x} + \dfrac{x - 3}{x^2 + x + 1}$

9. $-0.2, 0.3, 1.4$
10. Figure 68
11. Figure 69

Fig. 68

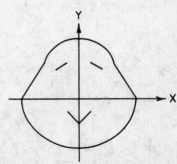

Fig. 69

Exercise IX.

2. a) $\log N = 3 \log x + \dfrac{1}{2} \log y$

b) $\log N = \log 3 + \dfrac{1}{2} \log 2.9$

c) $\log N = \log 32.3. + \dfrac{1}{2} \log 5.18 - \log 12 - 2 \log 4.9$

d) $\log N = \log g + 2 \log V - 3 \log t - \dfrac{1}{2} \log h$

3. a) 2.8073   **b)** 1.8927   **c)** 3

d) 0.3154   **e)** 1.3652   **f)** $\dfrac{1}{2}$

4. a) 2.04   **b)** -3 **c)** 0.9   **d)** $\pm \sqrt{2}$

e) $\dfrac{3}{2} \pm \dfrac{\sqrt{5}}{2}$ **f)** 0.69315   **g)** $\dfrac{5}{2} \pm \dfrac{\sqrt{345}}{6}$ **h)** 6, -7 **i)** $\dfrac{3}{175}$

5. a) $x \in R$ except $0 \le x \le \dfrac{1}{2}$ **b)** $x \in R$ except $-2 \le x \le 0$

c) $x > \dfrac{4}{3}$ **d)** $x \in R$ except $0 \le x \le 2$

6. a) $x > 2$ **b)** $x > 0$ **c)** $0 < x < 1$ **d)** $2 < x < 6$

8. $\cosh^{-1} x = \ln (x + \sqrt{x^2 - 1}), |x| > 1$

Exercise X.

1. a) $-5, -7, -9, \ldots$ **b)** $2, 5, 8, \ldots$ **c)** $189, 567, 1701, \ldots$

d) $16, -32, 64, \ldots$ **e)** $21, 25, 29$ **f)** $2, 1, \dfrac{1}{2}$

2. a) $l = 59, S = 279$ **b)** $n = 11, l = 63$ **c)** $d = 4, S = 0$

d) $l = \dfrac{3}{16}, S = \dfrac{381}{16}$ **e)** $n = 6, S = \dfrac{665}{36}$ **f)** $r = -5, 4; l = 75, 48$

g) $54, 45, 36, 27, 18$ **h)** $16, 128$

3. a) $2^n - 1$ **b)** n^2 **c)** $n(n + 1)$

4. a) 2   **b)** $\dfrac{2}{3}$ **c)** $\dfrac{3}{2}$ **d)** $\dfrac{5}{4}$ **e)** 18   **f)** $\dfrac{10}{9}$ **g)** $1.\overline{1}$

5. a) convergent   **b)** divergent   **c)** divergent   **d)** divergent
e) divergent   **f)** convergent   **g)** divergent

6. a) $x^7 - 14x^6 y^2 + 84x^5 y^4 - 280x^4 y^6 + \ldots$

b) $(3x)^{\frac{1}{3}} - \dfrac{2}{3}(3x)^{-\frac{2}{3}} y - \dfrac{4}{9}(3x)^{-\frac{5}{3}} y^2 - \dfrac{40}{81}(3x)^{-\frac{8}{3}} y^3 - \ldots$

c) $x^{-1} + 4x^{-\frac{3}{2}} y^{\frac{1}{3}} + 12x^{-2} y^{\frac{2}{3}} + 32x^{-\frac{5}{2}} y + \ldots$

d) $a^7 + 7a^6 bx + 21a^5 b^2 x^2 + 35a^4 b^3 x^3 + \ldots$

e) $1 + \frac{1}{2}x - \frac{1}{8}x^2 + \frac{1}{16}x^3 - \ldots$

f) $1 - \frac{1}{2}x + \frac{3}{8}x^2 - \frac{5}{16}x^3 + \ldots$

g) $\frac{1}{8} - \frac{9}{16}x + \frac{27}{16}x^2 - \frac{135}{32}x^3 + \ldots$

h) $1 - 2bx + 3b^2x^2 - 4b^3x^3 + \ldots$

7. a) $\frac{231}{1024}(3^6)(2x)^{-\frac{13}{2}}y^3$ b) $6435x^7y^8$

c) $-6a^{-7}b^5x^5$

8. a) $|x| < 1$, $|x| < 1$, $|x| < \frac{1}{6}$, $|x| < \frac{1}{b}$

b) $|x| < 1$ c) $|x^2| < 1$ d) all x e) $0 \le x \le 2$

9. $x + \frac{x^3}{3!} + \frac{x^5}{5!} + \frac{x^7}{7!} + \ldots$, all x

10. 0.866

Exercise XI.

1. a) $\{a, b, c\}, \{d\}, \emptyset, \{e, f\}$

b) $\{a, b\}, \{c, d\}, \{e, f\}, \emptyset$

2. Partition the set into 3 sets each of 3 elements

3. 7   4. 8

6. a) 3024   b) 990   c) 657720

7. a) 24   b) 64

8. 3780   9. 13824   10. 8

11. a) 9   b) 7

12. a) 792   b) 84   c) n

13. 120   14. 739741   15. 1722

16. a) 36   b) 216

17. $\frac{1}{36}, \frac{1}{18}, \frac{1}{12}, \frac{1}{9}, \frac{5}{36}, \frac{1}{6}, \frac{5}{36}, \frac{1}{9}, \frac{1}{12}, \frac{1}{18}, \frac{1}{36}$

18. a) 3   b) 6

19. a) $\frac{4}{81}$ b) $\frac{2}{45}$

20. a) $\frac{1}{81}$ b) $\frac{25}{1296}$ c) $\frac{1}{36}$ d) $\frac{5}{216}$

21. $\frac{1}{5}$ 22. a) $\frac{125}{3888}$ b) $\frac{23}{648}$

23. $\frac{115}{324} = \$0.35$ 24. \$0.50   25. $\frac{313}{729}$

Exercise XII.

1. Final columns only (note location of **h**)

a	b	c	d	e	f	g	i
F	F	T	F	T	F	T	T
T	F	T	F	T	F	T	T
T	F	F	F	T	F	T	
T	T		F	T	F	T	

h	j	k	l	m	n	o	p
T	T	T	T	T	F	T	T
F	F	T	T	T	T	T	F
T	F	T	T	T	T	T	T
F	T	T	T	T	T	F	F
T						T	F
T						T	F
F						T	T
F						T	F

2. All tautologies except **h**

3. **a)** See p. 209
   b) Converse: If $2a$ is divisible by 4, then a is an even integer.
   Inverse: If a is an odd integer, then $2a$ is not divisible by 4.
   Contrapositive: If $2a$ is not divisible by 4, then a is not an even integer.
   c) Check yourself.
   d) $q \rightarrow p, (\sim p) \rightarrow (\sim q), (\sim q) \rightarrow (\sim p)$

4. See p. 209

5. Hint: $(2k + 1) + (2r + 1) = 2(k + r + 1) = 2n$

6. $(1, 1, 0)$, The set $(1, 1, 1)$ does not satisfy the equation.

7. See p. 213

8. See illustration p. 215

10. Test the axioms. See p. 219

11. **a)** $(a \circ b) = c, c \circ c = e; (b \circ c) = a, a \circ a = e$
    c) Each is its own inverse.

12. Test the axioms.

13. There is no additive inverse.

14. Test the axioms.

15. **a)** Show that $\dfrac{n}{n + 1} + \dfrac{1}{(n + 1)(n + 2)} = \dfrac{n + 1}{n + 2}$

ANSWERS TO THE EXAMINATIONS

Part I.
Test #1.

1. a) -4
   b) $4a^4x^3y^2$
   c) $3x - y$
   d) $\frac{8}{3}xy$
   e) $9y^2 + 6xy + 4x^2$
   f) $2x + 3\sqrt{xy} - 2y$
   g) $\frac{1}{13}(5 + 12i)$
   h) $\frac{1}{x - 3}$
   i) $1 - 3i$
   j) $\frac{8x + 1}{(2x - 3)(x + 2)(x - 1)}$

2. a) $\frac{8}{7}$
   b) 0
   c) $\frac{1}{17}$
   d) $\frac{7b}{a - 3}$
   e) 2

3. 3 and 35

Test #2.

1. a) $[1, 7]$
   b) $\{1, 2, 3, 4, 5, 6, 7\}$
   c) $[2, 5]$
   d) $\{2, 3\}$
   e) $\emptyset$
   f) U

2. Figure 70

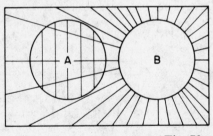 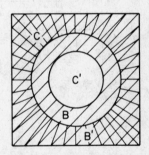

Fig. 70

3. a) $\{6, 8\}$
   b) U
   c) $\{2, 4, 8\}$
   d) $\{6, 8\}$
   e) $\{4, 6\}$
4. a) $\{(1, 2), (1, 4), (3, 2), (3, 4)\}$
   b) $\{(x, a), (x, b), (y, a), (y, b)\}$
   c) $\{(1, x), (1, y), (2, x), (2, y), (3, x), (3, y)\}$
5. $1 \leftrightarrow -1, 2 \leftrightarrow -2, 3 \leftrightarrow -3, \ldots n \leftrightarrow -n, \ldots$

Test #3.

1. $-1, 1, -3, \frac{1}{3}, 6, \frac{1}{6}, -\sqrt{2}, \frac{\sqrt{2}}{2}, \frac{1}{2}, 2$
2. See page 54
3. $N = \frac{5}{3}$

251

4. $[1, 6]$

5. a) $2 + 7i$ **b)** -10 **c)** $\dfrac{5 + 9i}{4}$

Test #4.

1.

	Value at			
Function	-1	0	1	3
a) $f(x) = x^2 - 3$	-2	-3	-2	6
b) $g(x) = x - 1$	-2	-1	0	2
c) $f + g = x^2 + x - 4$	-4	-4	-2	8
d) $f - g = x^2 - x - 2$	0	-2	-2	4
e) $fg = x^3 - x^2 - 3x + 3$	4	3	0	12
f) $\dfrac{f}{g} = \dfrac{x^2 - 3}{x - 1}$	1	3	—	3
g) $f(g) = x^2 - 2x - 2$	1	-2	-3	1

2. Values:

x	-3	-2	-1	0	1	2	3
y	6	4	2	0	2	4	6

3. a) $y = \dfrac{1}{3}(x + 2)$ **b)** $y = \dfrac{1}{2}(3 - x)$

    c) $y = \dfrac{1}{2}x$ **d)** $y = x - 1, x < 1$

4. 3

Test #5.

1. $m = \dfrac{2}{3}, b = 4$

2. $5y - 2x + 11 = 0, m = \dfrac{2}{5}$

3. $x = -1, y = -3$

4. $\begin{pmatrix} -2 & 2 \\ 10 & 14 \\ 24 & 18 \end{pmatrix}$

5. a) $x \geq 4$

    b) Figure 71

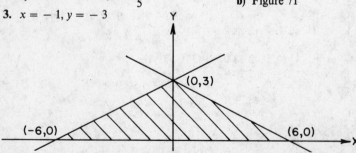

Fig. 71

Test #6.

1. a) $1 \pm \frac{2}{3}\sqrt{3}$ b) $-\frac{3}{4} \pm \frac{\sqrt{23}}{4}i$

2. $(2, 1)$ and $(2, -1)$ 5. $-3 < x < 2$

3. $y^2 = 4 - x, -12 \le x \le 4$

4. $1, -\frac{1}{2} \pm \frac{\sqrt{3}}{2}i$ 6. $1 + \frac{2}{x-1} + \frac{2}{(x-1)^2}$

7. Figure 72

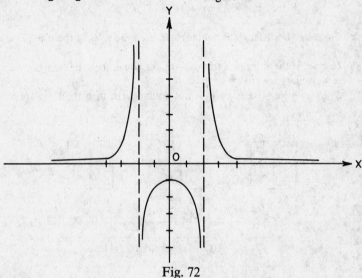

Fig. 72

Test #7.

1. $\log g + \frac{3}{2} \log h - 2 \log t - \log v$ 3. $y = \ln (x + \sqrt{1 + x^2})$

4. $x > 2$

2. 1.7712

Test #8.

1. $\frac{26}{7}, \frac{31}{7}, \frac{36}{7}, \frac{41}{7}, \frac{46}{7}, \frac{51}{7}$

2. 4, 8, 16, 32

3. $\sqrt{2x} - \frac{3}{2}(2x)^{-\frac{1}{2}} - \frac{9}{8}(2x)^{-\frac{3}{2}} - \frac{27}{16}(2x)^{-\frac{5}{2}}$

4. all x

5. $\cosh x = 1 + \frac{x^2}{2!} + \frac{x^4}{4!} + \ldots$

Test #9.

1. $n = 12$ 2. $\frac{3}{17}$ 3. 24

4. 6 5. $\{1, 2, 3\}, \{4\}, \emptyset, \{5, 6\}$

Test #10.

1.

p	q	$p \to q$	$q \to p$	$(3) \wedge (4)$
T	T	T	T	T
T	F	F	T	F
F	T	T	F	F
F	F	T	T	T

2. *Converse:* If a triangle is equiangular, then the sides of the triangle are all equal.

Inverse: If the sides of a triangle are not equal, then the triangle is not equiangular.

Contrapositive: If a triangle is not equiangular, then the sides are not equal.

3. Figure 73

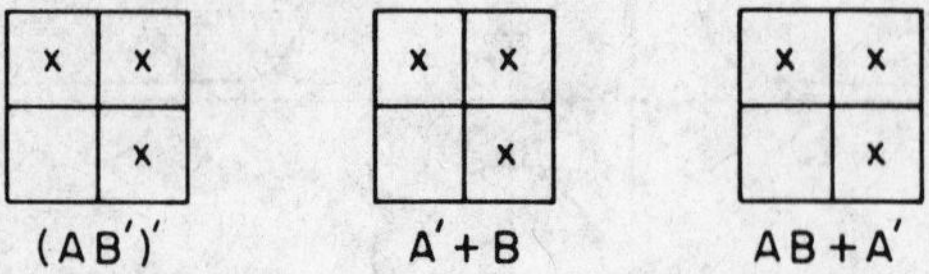

$(AB')'$ $A' + B$ $AB + A'$

Fig. 73

4.

x	y	$x + y$	$x(x + y)$
0	0	0	0
0	1	1	0
1	0	1	1
1	1	1	1

5. If not-p, then not-q.

Part II.

B. 1. a) $-\dfrac{bc}{a^2}$ **b)** $k = -\dfrac{25}{4}, \left(x = \dfrac{7}{2}, y = \dfrac{3}{4}\right)$

2. a) $x^4 + 6x^3 + 11x^2 + 6x + 1 = (x^2 + 3x + 1)^2$

3. a) $x \le -3, x \ge 5$ **b)** $\sqrt{ab} < \dfrac{a + b}{2}$

4. a) Rationalize the denominator.   **b)** 300

5. $8, -4, 2, 8$

6. 18 mi.

C. The proofs may be found in the text.

D.

1	2	3	4	5	6	7	8	9	10	11	12
T	T	F	T	F	F	F	T	T	T	T	T

13	14	15	16	17	18	19	20	21	22	23	24
F	T	T	F	T	T	T	T	F	T	T	T

25	26	27	28	29	30	31	32	33	34	35	36
F	T	T	T	T	T	T	F	F	T	F	F

37	38	39	40	41	42	43	44	45	46	47	48
F	T	T	T	T	T	T	T	T	T	T	F

49	50
F	T

E. 1. $N = 0.1033$

2. $-\dfrac{1}{2} \pm \dfrac{\sqrt{3}}{2} i$

3. 0.8156

4. 5.045

5. 1.110

6. $2\pi - \sqrt{3} \doteq 4.5511$

7. 3.00417

8. $2, -2, \dfrac{1}{2}, -\dfrac{2}{3}$

9. $\dfrac{32}{11}, \dfrac{17}{11}, \dfrac{16}{11}$

10. $\left(\dfrac{1}{3}\sqrt{6}, \dfrac{1}{2}\sqrt{6}\right), \left(-\dfrac{1}{3}\sqrt{6}, -\dfrac{1}{2}\sqrt{6}\right), \left(-\dfrac{1}{2}\sqrt{6}, \dfrac{1}{2}\sqrt{6}\right), \left(\dfrac{1}{2}\sqrt{6}, -\dfrac{1}{2}\sqrt{6}\right)$

F. 1. $\dfrac{b^2}{a^2}$

2. $x < -4$ and $x > -3$

3. Hyperbola, center at $(1, 3)$, vertices at $(1, 0)$ and $(1, 6)$

4. Vertex at $(1, 2)$, axis parallel to X-axis, curve opens to the right

5. $f^{-1}: y = -2\sqrt{4 - x^2}$, domain $0 \le x \le 2$

6. 7 and 24

7. all x, $x \neq 0$

8. $D_x y = -\dfrac{x}{y}$

9. ± 5

10. Two straight lines defined by $2x - 3y = 0$ and $3x + y = 2$

G. The definitions may be found in the text. Check the index for the corresponding pages.

H.

No.	Description	Example
1.	implies	$x - 3 = 0 \rightarrow x = 3$
2.	exclusive disjunction	a number is even or odd
3.	incremental change	$x_1 = 3, x_2 = 3.1$; then $\Delta x = 0.1$
4.	absolute value	$|-3| = 3$
5.	ordered pair	$(2, 7)$
6.	permutation	$P(5, 2) = 5 \cdot 4 = 20$
7.	plus or minus	$\sqrt{4} = \pm 2$
8.	greater than	$7 > 5$
9.	square root of n	$\sqrt{16} = 4$
10.	equivalence	$x - 2 = 7 \leftrightarrow x = 9$
11.	union of two sets	$[2, 5] \cup [3, 6] = [2, 6]$
12.	null set	$[1, 2] \cap [5, 7] = \emptyset$
13.	a set	$R = \{x \mid x \text{ is a real number}\}$
14.	complement	$U = [2, 5], A = [2, 3]$, then $A' = \,]3, 5]$
15.	element of a set	If $A = \{5, 7, 9\}$, then $5 \in A$.
16.	repeating decimal	$1.\overline{3} = 1.3333\ldots$
17.	negative of	The negative of 5 is -5.
18.	value of a function	$f(x) = x^2 + 1$
19.	not-p	$p : n$ is positive, $\sim p : n$ is not positive
20.	a function	$f : y = 2x + 3$
21.	largest integer	$\left[\dfrac{7}{2}\right] = 3$
22.	inverse of a function	$f : y = 2x + 3, f^{-1} : y = \dfrac{1}{2}(x - 3)$
23.	derivative	$y = 2x, D_x y = 2$
24.	factorial	$3! = 3 \cdot 2 \cdot 1 = 6$
25.	natural logarithm	$\ln 3$
26.	finite sum	$\sum\limits_{i=1}^{2} x^i = x + x^2$
27.	matrix	$\begin{pmatrix} 2 & -1 \\ 3 & 5 \end{pmatrix}$
28.	determinant	$\begin{vmatrix} 2 & -1 \\ 3 & 5 \end{vmatrix} = 13$
29.	hyperbolic sine	$\sinh 2$
30.	limit	$\lim\limits_{n \to \infty} \left(\dfrac{1}{n}\right) = 0$

I.

1. $\{1, 2, 3, 4, 5\}$	2. $\emptyset$	3. $[7, 12[$
4. A	5. 4	6. p
7. $\dfrac{4}{13} + \dfrac{19}{13}i$	8. A	9. -56

10. $8 + 8i$ **11.** $A - B$ **12.** $\dfrac{16}{11}$

13. -37 **14.** $\begin{pmatrix} 6 \\ 10 \\ 4 \end{pmatrix}$ **15.** $\dfrac{1}{5}\begin{pmatrix} 1 & 1 \\ -2 & 3 \end{pmatrix}$

16. A **17.** $-\dfrac{7}{3}$ **18.** $1, -\dfrac{1}{2} \pm \dfrac{\sqrt{3}}{2} i$

19. $-3, -4, 5$ **20.** 210

J. **1.** Perform the necessary matrix multiplications.

2. Replace $i, j, k,$ and I by their matrices and perform the matrix addition.

3. Show that $\bar{y}_1 x_2 + \bar{x}_1 \bar{y}_2 = \overline{x_1 y_2 + y_1 \bar{x}_2}$ and $\bar{x}_1 \bar{x}_2 - \bar{y}_1 y_2 =$

$= \overline{x_1 x_2 - y_1 \bar{y}_2}.$

4. Replace $I, i, j,$ and k by their matrices, perform the addition, and use the definitions of x and y.

5. Perform the matrix multiplication $q\bar{q}$.

6. Obtain the matrix representation of $\overline{q_1 q_2}$ and $\bar{q}_2 \bar{q}_1$; then show

$\overline{x_1 x_2 - y_1 \bar{y}_2} = \bar{x}_2 \bar{x}_1 - y_2 \bar{y}_1$ and $\overline{x_1 y_2 + y_1 \bar{x}_2} = \bar{x}_1 \bar{y}_2 + x_2 \bar{y}_1.$

7. Perform the matrix multiplication.

8. Show that $q_1 + q_2 = q_2 + q_1.$

9. Show that $q_1 q_2 \neq q_2 q_1.$

10. Show that quaternions satisfy the Closure, Associative, Identity, Inverse, and Commutative Axioms for addition.

Part III.
Exam. #1. Part A.

1	2	3	4	5	6	7	8	9	10	11	12
T	F	F	F	F	F	T	T	T	T	T	T

13	14	15	16	17	18	19	20	21	22	23	24
T	T	T	F	F	T	T	T	F	F	T	F

25
T

Part B.

1. $2, \pm \sqrt[4]{2}, \pm \sqrt[4]{2}\, i$

2. $f^{-1}: y = \sqrt{-x}, -9 \leq x \leq 0$

3. $\dfrac{7}{32}$ 5. $-7 < x < 2$

6. See Figures 62 and 63, p. 217.

7. $f(1, 5) = -5$ (min.), $f(5, 1) = 11$ (max.).

8. **a)** $[-1, 7]$ **b)** $\{2, 6\}$ **c)** $\{6, 7, 8, 9, 10\}$
   d) $\{(2, 0), (2, 1), (4, 0), (4, 1)\}$ **e)** A

9. $2, 0, 1, -2$

10. $1 - x - \dfrac{1}{2}x^2 - \dfrac{1}{2}x^3 - \dfrac{5}{8}x^4 - \dfrac{7}{8}x^5 - \cdots$

Exam. #2. Part A.

1	2	3	4	5	6	7	8	9	10	11	12	13
F	F	F	T	F	F	F	T	F	T	T	T	T

14	15	16	17	18	19	20	21	22	23	24	25
T	F	T	T	F	T	T	T	T	T	T	T

Part B.

1. 12   2. $-3, 7, \dfrac{1}{6}(-1 \pm \sqrt{13})$

3. $1 + x - \dfrac{1}{2}x^2 + \dfrac{1}{2}x^3 - \dfrac{5}{8}x^4 + \cdots$

4. $\dfrac{1}{y^2} + \dfrac{1}{(y-1)^2}$ 6. Figure 26   7. Figure 35

8. $\dfrac{113}{9}$ 9. Use mathematical induction.   10. See page 53.

Exam. #3.

1. See page 48.   2. $-8i$ 3. $(5, 6, 6)$

4. $\dfrac{1}{72}$ 5. See page 134.   6. $\log \sqrt[3]{4}$

7. 0.71   8. **a)** $(3, 1)$, **b)** $\{1, 3, 5, 7\}$

9. See page 95.   10. $y = -\sqrt{16 - x^2}, d_{f^{-1}} : 0 \le x \le 4$

TABLES

APPENDIX

Table 1. Powers & Roots

n	n^2	n^3	$\sqrt{n}$	$\sqrt{10n}$	$\sqrt[3]{n}$
1	1	1	1.00000	3.16228	1.00000
2	4	8	1.41421	4.47214	1.25992
3	9	27	1.73205	5.47723	1.44225
4	16	64	2.00000	6.32456	1.58740
5	25	125	2.23607	7.07107	1.70998
6	36	216	2.44949	7.74597	1.81712
7	49	343	2.64575	8.36660	1.91293
8	64	512	2.82843	8.94427	2.00000
9	81	729	3.00000	9.48683	2.08008
10	100	1000	3.16228	10.00000	2.15444
11	121	1331	3.31662	10.48809	2.22398
12	144	1728	3.46410	10.95445	2.28943
13	169	2197	3.60555	11.40175	2.35134
14	196	2744	3.74166	11.83216	2.41014
15	225	3375	3.87298	12.24745	2.46621
16	256	4096	4.00000	12.64911	2.51984
17	289	4913	4.12311	13.03840	2.57128
18	324	5832	4.24264	13.41641	2.62074
19	361	6859	4.35890	13.78405	2.66840
20	400	8000	4.47214	14.14214	2.71442
21	441	9261	4.58258	14.49138	2.75892
22	484	10648	4.69042	14.83240	2.80204
23	529	12167	4.79583	15.16575	2.84387
24	576	13824	4.89898	15.49193	2.88450
25	625	15625	5.00000	15.81139	2.92402
26	676	17576	5.09902	16.12452	2.96250
27	729	19683	5.19615	16.43168	3.00000
28	784	21952	5.29150	16.73320	3.03659
29	841	24389	5.38516	17.02939	3.07232
30	900	27000	5.47723	17.32051	3.10723
31	961	29791	5.56776	17.60682	3.14138
32	1024	32768	5.65685	17.88854	3.17480
33	1089	35937	5.74456	18.16590	3.20753
34	1156	39304	5.83095	18.43909	3.23961
35	1225	42875	5.91608	18.70829	3.27107
36	1296	46656	6.00000	18.97367	3.30193
37	1369	50653	6.08276	19.23538	3.33222
38	1444	54872	6.16441	19.49359	3.36198
39	1521	59319	6.24500	19.74842	3.39121
40	1600	64000	6.32456	20.00000	3.41995
41	1681	68921	6.40312	20.24846	3.44822
42	1764	74088	6.48074	20.49390	3.47603
43	1849	79507	6.55744	20.73644	3.50340
44	1936	85184	6.63325	20.97618	3.53035
45	2025	91125	6.70820	21.21320	3.55689
46	2116	97336	6.78233	21.44761	3.58305
47	2209	103823	6.85566	21.67948	3.60883
48	2304	110592	6.92820	21.90890	3.63424
49	2401	117649	7.00000	22.13594	3.65931
50	2500	125000	7.07107	22.36068	3.68403

Table 1. Continued

n	n²	n³	√n	√10n	∛n
51	2601	132651	7.14143	22.56318	3.70843
52	2704	140608	7.21110	22.80351	3.73251
53	2809	148877	7.28011	23.02173	3.75629
54	2916	157464	7.34847	23.23790	3.77976
55	3025	166375	7.41620	23.45208	3.80295
56	3136	175616	7.48332	23.66432	3.82586
57	3249	185193	7.54983	23.87467	3.84850
58	3364	195112	7.61577	24.08319	3.87088
59	3481	205379	7.68115	24.28992	3.89300
60	3600	216000	7.74597	24.49490	3.91487
61	3721	226981	7.81025	24.69818	3.93650
62	3844	238328	7.87401	24.89980	3.95789
63	3969	250047	7.93725	25.09980	3.97906
64	4096	262144	8.00000	25.29822	4.00000
65	4225	274625	8.06226	25.49510	4.02073
66	4356	287496	8.12404	25.69047	4.04124
67	4489	300763	8.18535	25.88436	4.06155
68	4624	314432	8.24621	26.07681	4.08166
69	4761	328509	8.30662	26.26785	4.10157
70	4900	343000	8.36660	26.45751	4.12128
71	5041	357911	8.42615	26.64583	4.14082
72	5184	373248	8.48528	26.83282	4.16017
73	5329	389017	8.54400	27.01851	4.17934
74	5476	405224	8.60233	27.20294	4.19834
75	5625	421875	8.66025	27.38613	4.21716
76	5776	438976	8.71780	27.56810	4.23582
77	5929	456533	8.77496	27.74887	4.25432
78	6084	474552	8.83176	27.92848	4.27266
79	6241	493039	8.88819	28.10694	4.29084
80	6400	512000	8.94427	28.28427	4.30887
81	6561	531441	9.00000	28.46050	4.32675
82	6724	551368	9.05539	28.63564	4.34448
83	6889	571787	9.11043	28.80972	4.36207
84	7056	592704	9.16515	28.98275	4.37952
85	7225	614125	9.21954	29.15476	4.39683
86	7396	636056	9.27362	29.32576	4.41400
87	7569	658503	9.32738	29.49576	4.43105
88	7744	681472	9.38083	29.66479	4.44796
89	7921	704969	9.43398	29.83287	4.46474
90	8100	729000	9.48683	30.00000	4.48140
91	8281	753571	9.53939	30.16621	4.49794
92	8464	778688	9.59166	30.33150	4.51436
93	8649	804357	9.64365	30.49590	4.53066
94	8836	830584	9.69536	30.65942	4.54684
95	9025	857375	9.74679	30.82207	4.56290
96	9216	884736	9.79796	30.98387	4.57886
97	9409	912673	9.84886	31.14482	4.59470
98	9604	941192	9.89949	31.30495	4.61044
99	9801	970299	9.94987	31.46427	4.62606
100	10000	1000000	10.00000	31.62278	4.64159

Table 2. Four-Place Common Logarithms

N	0	1	2	3	4	5	6	7	8	9
10	0000	0043	0086	0128	0170	0212	0253	0294	0334	0374
11	0414	0453	0492	0531	0569	0607	0645	0682	0719	0755
12	0792	0828	0864	0899	0934	0969	1004	1038	1072	1106
13	1139	1173	1206	1239	1271	1303	1335	1367	1399	1430
14	1461	1492	1523	1553	1584	1614	1644	1673	1703	1732
15	1761	1790	1818	1847	1875	1903	1931	1959	1987	2014
16	2041	2068	2095	2122	2148	2175	2201	2227	2253	2279
17	2304	2330	2355	2380	2405	2430	2455	2480	2504	2529
18	2553	2577	2601	2625	2648	2672	2695	2718	2742	2765
19	2788	2810	2833	2856	2878	2900	2923	2945	2967	2989
20	3010	3032	3054	3075	3096	3118	3139	3160	3181	3201
21	3222	3243	3263	3284	3304	3324	3345	3365	3385	3404
22	3424	3444	3464	3483	3502	3522	3541	3560	3579	3598
23	3617	3636	3655	3674	3692	3711	3729	3747	3766	3784
24	3802	3820	3838	3856	3874	3892	3909	3927	3945	3962
25	3979	3997	4014	4031	4048	4065	4082	4099	4116	4133
26	4150	4166	4183	4200	4216	4232	4249	4265	4281	4298
27	4314	4330	4346	4362	4378	4393	4409	4425	4440	4456
28	4472	4487	4502	4518	4533	4548	4564	4579	4594	4609
29	4624	4639	4654	4669	4683	4698	4713	4728	4742	4757
30	4771	4786	4800	4814	4829	4843	4857	4871	4886	4900
31	4914	4928	4942	4955	4969	4983	4997	5011	5024	5038
32	5051	5065	5079	5092	5105	5119	5132	5145	5159	5172
33	5185	5198	5211	5224	5237	5250	5263	5276	5289	5302
34	5315	5328	5340	5353	5366	5378	5391	5403	5416	5428
35	5441	5453	5465	5478	5490	5502	5514	5527	5539	5551
36	5563	5575	5587	5599	5611	5623	5635	5647	5658	5670
37	5682	5694	5705	5717	5729	5740	5752	5763	5775	5786
38	5798	5809	5821	5832	5843	5855	5866	5877	5888	5899
39	5911	5922	5933	5944	5955	5966	5977	5988	5999	6010
40	6021	6031	6042	6053	6064	6075	6085	6096	6107	6117
41	6128	6138	6149	6160	6170	6180	6191	6201	6212	6222
42	6232	6243	6253	6263	6274	6284	6294	6304	6314	6325
43	6335	6345	6355	6365	6375	6385	6395	6405	6415	6425
44	6435	6444	6454	6464	6474	6484	6493	6503	6513	6522
45	6532	6542	6551	6561	6571	6580	6590	6599	6609	6618
46	6628	6637	6646	6656	6665	6675	6684	6693	6702	6712
47	6721	6730	6739	6749	6758	6767	6776	6785	6794	6803
48	6812	6821	6830	6839	6848	6857	6866	6875	6884	6893
49	6902	6911	6920	6928	6937	6946	6955	6964	6972	6981
50	6990	6998	7007	7016	7024	7033	7042	7050	7059	7067
51	7076	7084	7093	7101	7110	7118	7126	7135	7143	7152
52	7160	7168	7177	7185	7193	7202	7210	7218	7226	7235
53	7243	7251	7259	7267	7275	7284	7292	7300	7308	7316
54	7324	7332	7340	7348	7356	7364	7372	7380	7388	7396

Table 2. Continued

N	0	1	2	3	4	5	6	7	8	9
55	7404	7412	7419	7427	7435	7443	7451	7459	7466	7474
56	7482	7490	7497	7505	7513	7520	7528	7536	7543	7551
57	7559	7566	7574	7582	7589	7597	7604	7612	7619	7627
58	7634	7642	7649	7657	7664	7672	7679	7686	7694	7701
59	7709	7716	7723	7731	7738	7745	7752	7760	7767	7774
60	7782	7789	7796	7803	7810	7818	7825	7832	7839	7846
61	7853	7860	7868	7875	7882	7889	7896	7903	7910	7917
62	7924	7931	7938	7945	7952	7959	7966	7973	7980	7987
63	7993	8000	8007	8014	8021	8028	8035	8041	8048	8055
64	8062	8069	8075	8082	8089	8096	8102	8109	8116	8122
65	8129	8136	8142	8149	8156	8162	8169	8176	8182	8189
66	8195	8202	8209	8215	8222	8228	8235	8241	8248	8254
67	8261	8267	8274	8280	8287	8293	8299	8306	8312	8319
68	8325	8331	8338	8344	8351	8357	8363	8370	8376	8382
69	8388	8395	8401	8407	8414	8420	8426	8432	8439	8445
70	8451	8457	8463	8470	8476	8482	8488	8494	8500	8506
71	8513	8519	8525	8531	8537	8543	8549	8555	8561	8567
72	8573	8579	8585	8591	8597	8603	8609	8615	8621	8627
73	8633	8639	8645	8651	8657	8663	8669	8675	8681	8686
74	8692	8698	8704	8710	8716	8722	8727	8733	8739	8745
75	8751	8756	8762	8768	8774	8779	8785	8791	8797	8802
76	8808	8814	8820	8825	8831	8837	8842	8848	8854	8859
77	8865	8871	8876	8882	8887	8893	8899	8904	8910	8915
78	8921	8927	8932	8938	8943	8949	8954	8960	8965	8971
79	8976	8982	8987	8993	8998	9004	9009	9015	9020	9025
80	9031	9036	9042	9047	9053	9058	9063	9069	9074	9079
81	9085	9090	9096	9101	9106	9112	9117	9122	9128	9133
82	9138	9143	9149	9154	9159	9165	9170	9175	9180	9186
83	9191	9196	9201	9206	9212	9217	9222	9227	9232	9238
84	9243	9248	9253	9258	9263	9269	9274	9279	9284	9289
85	9294	9299	9304	9309	9315	9320	9325	9330	9335	9340
86	9345	9350	9355	9360	9365	9370	9375	9380	9385	9390
87	9395	9400	9405	9410	9415	9420	9425	9430	9435	9440
88	9445	9450	9455	9460	9465	9469	9474	9479	9484	9489
89	9494	9499	9504	9509	9513	9518	9523	9528	9533	9538
90	9542	9547	9552	9557	9562	9566	9571	9576	9581	9586
91	9590	9595	9600	9605	9609	9614	9619	9624	9628	9633
92	9638	9643	9647	9652	9657	9661	9666	9671	9675	9680
93	9685	9689	9694	9699	9703	9708	9713	9717	9722	9727
94	9731	9736	9741	9745	9750	9754	9759	9763	9768	9773
95	9777	9782	9786	9791	9795	9800	9805	9809	9814	9818
96	9823	9827	9832	9836	9841	9845	9850	9854	9859	9863
97	9868	9872	9877	9881	9886	9890	9894	9899	9903	9908
98	9912	9917	9921	9926	9930	9934	9939	9943	9948	9952
99	9956	9961	9965	9969	9974	9978	9983	9987	9991	9996

APPENDIX

Table 3. ln x, e^x, e^{-x}, sinh x, cash x

x	ln x	e^x	e^{-x}	sinh x	cosh x
0.0	....	1.00000	1.00000	0.0000	1.0000
0.1	-2.30258	1.10517	$9.04837-1$	0.1002	1.0050
0.2	-1.60943	1.22140	$8.18731-1$	0.2013	1.0201
0.3	-1.20397	1.34986	$7.40818-1$	0.3045	1.0453
0.4	-0.91629	1.49182	$6.70320-1$	0.4108	1.0811
0.5	-0.69314	1.64872	$6.06531-1$	0.5211	1.1276
0.6	-0.51082	1.82212	$5.48812-1$	0.6367	1.1855
0.7	-0.35667	2.01375	$4.96585-1$	0.7586	1.2552
0.8	-0.22314	2.22554	$4.49329-1$	0.8881	1.3374
0.9	-0.10536	2.45960	$4.06570-1$	1.0265	1.4331
1.0	0.00000	2.71828	$3.67879-1$	1.1752	1.5431
1.1	0.09531	3.00417	$3.32871-1$	1.3356	1.6685
1.2	0.18232	3.32012	$3.01194-1$	1.5095	1.8107
1.3	0.26236	3.66930	$2.72532-1$	1.6984	1.9709
1.4	0.33647	4.05520	$2.46597-1$	1.9043	2.1509
1.5	0.40547	4.48169	$2.23130-1$	2.1293	2.3524
1.6	0.47000	4.95303	$2.01897-1$	2.3756	2.5775
1.7	0.53063	5.47395	$1.82684-1$	2.6456	2.8283
1.8	0.58779	6.04965	$1.65299-1$	2.9422	3.1075
1.9	0.64185	6.68589	$1.49569-1$	3.2682	3.4177
2.0	0.69315	7.38906	$1.35335-1$	3.6269	3.7622
2.1	0.74194	8.16617	$1.22456-1$	4.0219	4.1443
2.2	0.78846	9.02501	$1.10803-1$	4.4571	4.5679
2.3	0.83291	9.97418	$1.00259-1$	4.9370	5.0372
2.4	0.87547	11.0232	$9.07180-2$	5.4662	5.5569
2.5	0.91629	12.1825	$8.20850-2$	6.0502	6.1323
2.6	0.95551	13.4637	$7.42736-2$	6.6947	6.7690
2.7	0.99325	14.8797	$6.72055-2$	7.4063	7.4735
2.8	1.02962	16.4446	$6.08101-2$	8.1919	8.2527
2.9	1.06471	18.1741	$5.50232-2$	9.0596	9.1146
3.0	1.09861	20.0855	$4.97871-2$	10.018	10.068
3.1	1.13140	22.1980	$4.50492-2$	11.076	11.122
3.2	1.16315	24.5325	$4.07622-2$	12.246	12.287
3.3	1.19392	27.1126	$3.68832-2$	13.538	13.575
3.4	1.22378	29.9641	$3.33733-2$	14.965	14.999
3.5	1.25276	33.1155	$3.01974-2$	16.543	16.573
3.6	1.28093	36.5982	$2.73237-2$	18.286	18.313
3.7	1.30833	40.4473	$2.47235-2$	20.211	20.236
3.8	1.33500	44.7012	$2.23708-2$	22.339	22.362
3.9	1.36098	49.4024	$2.02419-2$	24.691	24.711
4.0	1.38629	54.5982	$1.83156-2$	27.290	27.308
4.1	1.41099	60.3403	$1.65727-2$	30.162	30.178
4.2	1.43508	66.6863	$1.49956-2$	33.336	33.351
4.3	1.45861	73.6998	$1.35686-2$	36.843	36.857
4.4	1.48160	81.4509	$1.22773-2$	40.719	40.732
4.5	1.50408	90.0171	$1.11090-2$	45.003	45.014
4.6	1.52606	99.4843	$1.00518-2$	49.737	49.747
4.7	1.54756	109.947	$9.09528-3$	54.969	54.978
4.8	1.56862	121.510	$8.22975-3$	60.751	60.759
4.9	1.58924	134.290	$7.44658-3$	67.141	67.149
x	ln x	e^x	e^{-x}	sinh x	cosh x

Table 3. Continued

x	$\ln x$	e^x	e^{-x}	$\sinh x$	$\cosh x$
5.0	1.60944	148.413	$6.73795-3$	74.203	74.210
5.1	1.62924	164.022	$6.09675-3$	82.008	82.014
5.2	1.64866	181.272	$5.51656-3$	90.633	90.639
5.3	1.66771	200.337	$4.99159-3$	100.17	100.17
5.4	1.68640	221.406	$4.51658-3$	110.70	110.71
5.5	1.70475	244.692	$4.08677-3$	122.34	122.35
5.6	1.72277	270.426	$3.69786-3$	135.21	135.22
5.7	1.74047	298.867	$3.34597-3$	149.43	149.44
5.8	1.75786	330.300	$3.02755-3$	165.15	165.15
5.9	1.77495	365.037	$2.73944-3$	182.52	182.52
6.0	1.79176	403.429	$2.47875-3$	201.71	201.72
6.1	1.80829	445.858	$2.24287-3$	222.93	222.93
6.2	1.82455	492.749	$2.02943-3$	246.37	246.38
6.3	1.84055	544.572	$1.83630-3$	272.29	272.29
6.4	1.85630	601.845	$1.66156-3$	300.92	300.92
6.5	1.87180	665.142	$1.50344-3$	332.57	332.57
6.6	1.88707	735.093	$1.36037-3$	367.55	367.55
6.7	1.90211	812.406	$1.23091-3$	406.20	406.20
6.8	1.91692	897.847	$1.11378-3$	448.92	448.92
6.9	1.93152	992.275	$1.00779-3$	496.14	496.14
7.0	1.94591	1096.63	$9.11882-4$	548.32	548.32
7.1	1.96009	1211.97	$8.25105-4$	605.98	605.98
7.2	1.97408	1339.43	$7.46586-4$	669.72	669.72
7.3	1.98787	1480.30	$6.75539-4$	740.15	740.15
7.4	2.00148	1635.98	$6.11253-4$	817.99	817.99
7.5	2.01490	1808.04	$5.53084-4$	904.02	904.02
7.6	2.02815	1998.20	$5.00451-4$	999.10	999.10
7.7	2.04122	2208.35	$4.52827-4$	1104.2	1104.2
7.8	2.05412	2440.60	$4.09735-4$	1220.3	1220.3
7.9	2.06686	2697.28	$3.70744-4$	1348.6	1348.6
8.0	2.07944	2980.96	$3.35463-4$	1490.5	1490.5
8.1	2.09186	3294.47	$3.03539-4$	1647.2	1647.2
8.2	2.10413	3640.95	$2.74654-4$	1820.5	1820.5
8.3	2.11626	4023.87	$2.48517-4$	2011.9	2011.9
8.4	2.12823	4447.07	$2.24867-4$	2223.5	2223.5
8.5	2.14007	4914.77	$2.03468-4$	2457.4	2457.4
8.6	2.15176	5431.66	$1.84106-4$	2715.8	2715.8
8.7	2.16332	6002.91	$1.66586-4$	3001.5	3001.5
8.8	2.17475	6634.24	$1.50733-4$	3317.1	3317.1
8.9	2.18605	7331.97	$1.36389-4$	3666.0	3666.0
9.0	2.19722	8103.08	$1.23410-4$	4051.5	4051.5
9.1	2.20827	8955.29	$1.11666-4$	4477.6	4477.6
9.2	2.21920	9897.13	$1.01039-4$	4948.6	4948.6
9.3	2.23001	10938.0	$9.14242-5$	5469.0	5469.0
9.4	2.24071	12088.4	$8.27241-5$	6044.2	6044.2
9.5	2.25129	13359.7	$7.48518-5$	6679.9	6679.9
9.6	2.26176	14764.8	$6.77287-5$	7382.4	7382.4
9.7	2.27213	16317.6	$6.12835-5$	8158.8	8158.8
9.8	2.28238	18033.7	$5.54516-5$	9016.9	9016.9
9.9	2.29253	19930.4	$5.01747-5$	9965.2	9965.2
x	$\ln x$	e^x	e^{-x}	$\sinh x$	$\cosh x$

Table 4. Constants

Values				Reciprocals		
π	3.14159	26535	89793	$\dfrac{1}{\pi}$	0.31830	98861	83791
$\dfrac{\pi}{2}$	1.57079	63267	94897	$\dfrac{2}{\pi}$	0.63661	97723	67582
2π	6.28318	53071	79586	$\dfrac{1}{2\pi}$	0.15915	49430	91895
π^2	9.86960	44010	89359	$\dfrac{1}{\pi^2}$	0.10132	11836	42338
$\sqrt{\pi}$	1.77245	38509	05516	$\dfrac{1}{\sqrt{\pi}}$	0.56418	95835	47756
$\sqrt{\dfrac{\pi}{2}}$	1.25331	41373	15500	$\sqrt{\dfrac{2}{\pi}}$	0.79788	45608	02865
$\sqrt{2\pi}$	2.50622	82746	31001	$\dfrac{1}{\sqrt{2\pi}}$	0.39894	22804	01433
e	2.71828	18284	59045	$\dfrac{1}{e}$	0.36787	94411	71442
e^2	7.38905	60989	30650	$\dfrac{1}{e^2}$	0.13533	52832	36613
$\sqrt{e}$	1.64872	12707	00128	$\dfrac{1}{\sqrt{e}}$	0.60653	06597	12633
$\log_{10} e$	0.43429	44819	03252	$\log_e 10$	2.30258	50929	94046

$$\begin{aligned}
\log_{10} \pi &= 0.49714 \quad 98726 \quad 94133 \\
1'' &= 0.00000 \quad 48481 \quad 36811 \quad 095 \quad \text{radians} \\
\sin 1'' &= 0.00000 \quad 48481 \quad 36811 \quad 076 \\
\tan 1'' &= 0.00000 \quad 48481 \quad 36811 \quad 133
\end{aligned}$$

INDEX